9/94

CRATERS, CAVERNS AND CANYONS

DELVING BENEATH THE EARTH'S SURFACE

The Changing Earth Series

JON ERICKSON

Facts On File

CRATERS, CAVERNS AND CANYONS:
DELVING BENEATH THE EARTH'S SURFACE

Facts On File, Inc.
460 Park Avenue South
New York NY 10016
USA

Library of Congress Cataloging-in-Publication Data
Erickson, Jon, 1948–
Craters, caverns and canyons : delving beneath the earth's surface / Jon Erickson.
p. cm. — (The changing earth series)
Includes bibliographical references and index.
ISBN 0-8160-2590-8
1. Geology, Structural. 2. Landforms. I. Title. II. Series: Erickson, Jon, 1948– Changing earth.
QE601.E73 1992
551.8—dc20 92–15626

A British CIP catalogue for this book is available from the British Library.

Text design by Ron Monteleone/Layout by Robert Yaffe
Jacket design by Catherine Hyman
Printed in the United States of America

10 9 8 7 6 5 4 3 2 1

This book is printed on acid-free paper.

CONTENTS

TABLES IN CRATERS, CAVERNS AND CANYONS

ACKNOWLEDGMENTS

The author thanks the following organizations for providing photographs for this book: the National Aeronautics and Space Administration (NASA), the National Oceanic and Atmospheric Administration (NOAA), the National Optical Astronomy Observatories (NOAO), the National Park Service, the U.S. Army, the U.S. Department of Agriculture–Forest Service, the U.S. Department of Energy, the U.S. Geological Survey (USGS) and the U.S. Navy.

INTRODUCTION

The surface of the Earth has been sculpted by a number of geologic processes. Throughout its long history, the Earth has been repeatedly bombarded by large meteorites, and occasionally asteroids the size of mountains struck the planet. The ancient impact craters they created are often difficult to locate because the planet's active geology has erased nearly all signs of them. Over a hundred remnants of ancient meteorite craters have been discovered scattered over the Earth, suggesting that it was just as heavily bombarded as other, less geologically active bodies in the Solar System where impacts are quite evident and numerous.

Here on Earth, groundwater has produced geologic forms as impressive as impact craters. Caves are some of the most spectacular examples of the handiwork of groundwater. Over geologic time, water has dissolved great quantities of soluble rock, forming extensive mazes of tunnels in the Earth's crust. A cave's elaborate architecture is determined by water seeping in from above, which creates many unusual formations and cave deposits of every description. Many of the same processes that form caves also create exquisite archways of rock that can span a ravine or valley. Running water, too, carving out solid rock, has produced some of the most magnificent scenery the planet has to offer. Perhaps nowhere is the power of erosion better demonstrated than the Grand Canyon of the Colorado River.

The world's oceans rival the continents for their rugged topography, and many canyons on the seafloor can hold several Grand Canyons. The oceans are crisscrossed by vast mountain ranges and deep canyons that have no comparison on land. Several times in the Earth's history the continents have been flooded by seas, and remnants of these ancient bodies of water can still be found. Nor are the seas permanent fixtures on the Earth's surface. The Mediterranean Sea appears to have dried out several times in the past, resulting in a huge empty pit. Regions where the Earth's crust is being stretched are collapsing, with some areas like California's Death Valley falling far below sea level.

Rifts slice through the ocean floor like gigantic cracks in the Earth, from which bleeds molten magma to manufacture new oceanic crust. This occurs at the midocean spreading ridges, where plates are diverging from each other. Continents rift when the crust is weakened by downdropped blocks of rock, bringing magma chambers near the surface where they further weaken the crust with extensive volcanic activity and violent earthquakes.

Similarly, earthquake faults scar the Earth's surface with downdropped blocks, escarpments and deep fissures. Large portions of the crust are bounded by long, parallel faults, forming a landscape of ridges and troughs. The vast majority of earthquakes coincides with the boundaries of these tectonic plates, which carry segments of crust around the globe. The most powerful earthquakes have faults that cut the Earth's surface. Thrust faults, which slide along a near horizontal surface, can be particularly damaging.

Entire sections of crust are gouged out by catastrophic ground failures. The Grand Canyon owes much of its existence to massive landslides that tore away whole canyon walls. Rock falls are spectacular, especially when large blocks fall straight down a vertical mountain face. Landslides on unstable slopes are particularly hazardous in mountainous and hilly areas. The earth can give way even on the gentlest slopes under favorable conditions. The weakening of sediment layers due to earthquakes can result in massive subsidence, causing the ground to suddenly collapse.

The making of the Earth's surface would not be complete without its numerous collapsed structures. Calderas form when the roof of a magma chamber collapses or when a powerful volcano decapitates itself, resulting in a broad depression that often fills with water to form a deep crater lake. The dissolution of subsurface materials or the withdrawal of underground fluids causes the surface to sink several feet. Other ground failures occur when subterranean sediments liquify during earthquakes or violent volcanic eruptions.

The removal of sediment by wind erosion scours the land, producing deflation basins and blowouts. The expulsion of gases under high pressure produces another type of blowout on land as well as on the seafloor. The Earth contains a variety of holes in the ground, including potholes, sinkholes, and numerous craters. Among the most unique depressions are kimberlite pipes, fumaroles and geysers, crater lakes, and lava lakes. These are just a few of the many wonders the Earth's active geology has to offer.

1

METEORITE IMPACTS

Throughout its long history, the Earth has been repeatedly bombarded by asteroids and comets, with a much higher incidence during the early years than recently. This is fortunate because the evolution of species would have turned out much differently if the impact rate had remained high throughout geologic history.

In its early development, the Earth was possibly struck by three Mars-size bodies, one of which might have created the moon. The gigantic impacts removed the Earth's primordial atmosphere, leaving it in a vacuum like the moon is today.

Sometimes asteroids the size of mountains struck the planet, inflicting a great deal of damage and causing the extinction of numerous species. Massive comet swarms, involving perhaps thousands of comets impacting all over the Earth, might also explain the disappearance of species. A popular theory for the extinction of the dinosaurs along with three quarters of all other known species contends that the Earth was hit by a large asteroid or comet nucleus that excavated a deep crater 100 miles or more wide and caused ecological chaos. Such a large depression should be easy to locate, yet it has eluded detection.

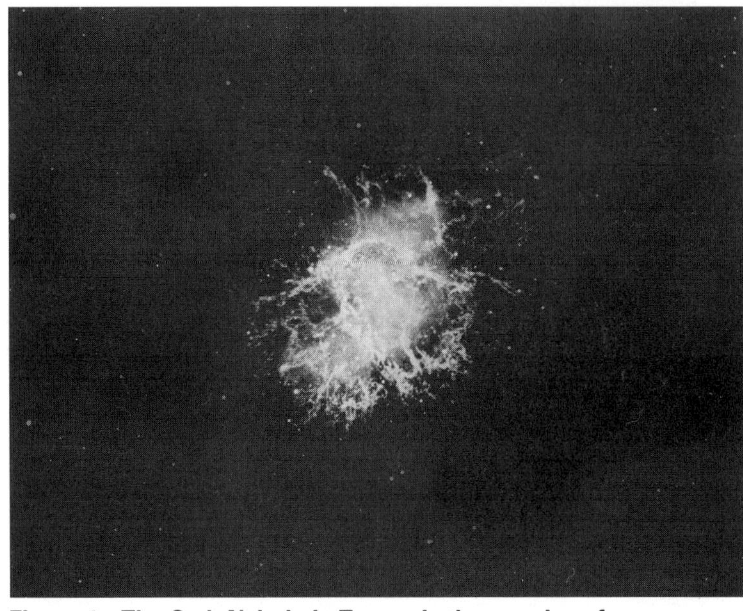

Figure 1 The Crab Nebula in Taurus is the remains of a supernova in 1054. Courtesy of NASA

PLANETARY ASSEMBLY

Around 4.6 billion years ago, the sun pulled itself together out of primordial gas and dust expelled by a supernova (Figure 1). The infant sun was surrounded by a protoplanetary disk composed of several bands of coarse particles, called planetesimals, which accreted from grains of dust and coalesced into larger bodies. Up to 100 trillion planetesimals orbited the sun during the Solar System's early stages of development. As they continued to grow, the small

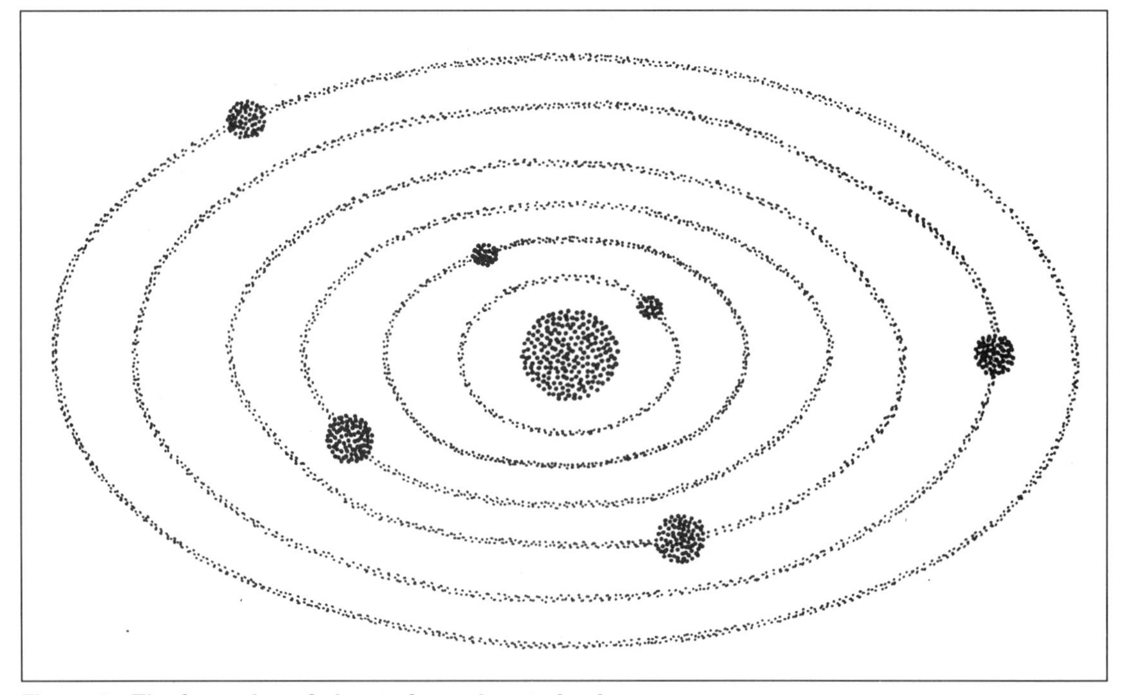

Figure 2 The formation of planets from planetesimals.

rocky chunks swung around the infant sun in highly elliptical orbits along the same plane (Figure 2). The constant collisions among planetesimals formed larger bodies, as much as 50 miles or more wide. These eventually became the planets that make up our Solar System.

When the sun ignited after reaching a certain critical mass, it cast off a strong solar wind of charged particles that was like a solar gale compared to the gentle breeze that blows today. It blew away the lighter components of the solar nebula and deposited them in the outer Solar System. The remaining planetesimals in the inner Solar System were composed mostly of stony and metallic fragments, ranging in size from fine sand grains to huge boulders, but most planetesimals were roughly the size of gravel. In the outer Solar System, where temperatures were much colder, rocky material along with solid chunks of water ice, frozen carbon dioxide, and crystalline methane and ammonia condensed.

The outer planets are believed to possess rocky cores about the size of the Earth, a mantle possibly composed of water ice and frozen methane, and a thick layer of compressed gas, mostly hydrogen and helium along with smaller amounts of methane and ammonia. Pluto, the moons of the outer planets, and the comets are essentially rock encased in a thick layer of ice or a jumble of rock and ice. Jupiter's composition is similar to the sun's, and if it had continued to grow it might have become hot enough to ignite into a small, brown dwarf star, resulting in a twin star system.

After about 10,000 years of formation, some planetary bodies grew to over 50 miles wide, but most of the planetary mass still resided with the small planetesimals. If not for the presence of a large gaseous medium in the solar nebula to slow down the planetesimals, the larger bodies would have continued to sweep up the remaining planetesimals, resulting in a Solar System composed of thousands of planetoids roughly 500 miles in diameter, about the size of the largest asteroids in existence today. This would result in a Solar System resembling Saturn and its rings (Figure 3).

Figure 3 Saturn from *Voyager 1* in November 1980. Courtesy of NASA

The Earth accreted into a homogenous mixture of silicates and iron-nickel by numerous meteorite impacts, and the entire process of planetary formation was completed in less than half a million years. As the Earth continued to grow, its orbit began to decay due to drag forces from leftover gases in interplanetary space. The formative planet slowly spiraled closer toward the sun, sweeping up additional planetesimals along its way. Eventually, the Earth's path around the sun was swept entirely clean of planetesimals, and its orbit stabilized near where it is today.

As the young planet cooled, it developed a thin basaltic crust, similar to that presently found on Venus, and indeed the moon and the inner planets offer clues to the Earth's early history. Among the few features common to the terrestrial planets was their ability to produce voluminous amounts of basaltic lavas. The Earth's original crust, which has long since disappeared, was unlike the basalts existing on the ocean floor. It was also quite distinct from modern continental crust, which first appeared around 4 billion years ago and represents less than half of 1 percent of the planet's total volume. Present-day plate tectonic processes could not have operated under these early hot conditions, with more vertical bubbling than horizontal sliding. Modern-style plate tectonics probably did not get fully underway much before 2.7 billion years ago.

Between 4.2 and 3.8 billion years ago, a massive meteorite shower, involving thousands of 50-mile-wide impactors, bombarded the Earth and its moon. The rest of the inner planets as well as the moons of the outer planets show numerous pockmarks from this invasion (Figure 4), with little happening since. The massive meteorite bombardment melted large portions of the Earth's crust by impact friction. Some 30 to 50 percent of the Earth's crust was con-

Figure 4 Saturn's heavily cratered moon Mimas from *Voyager 1* in November 1980. Courtesy of NASA

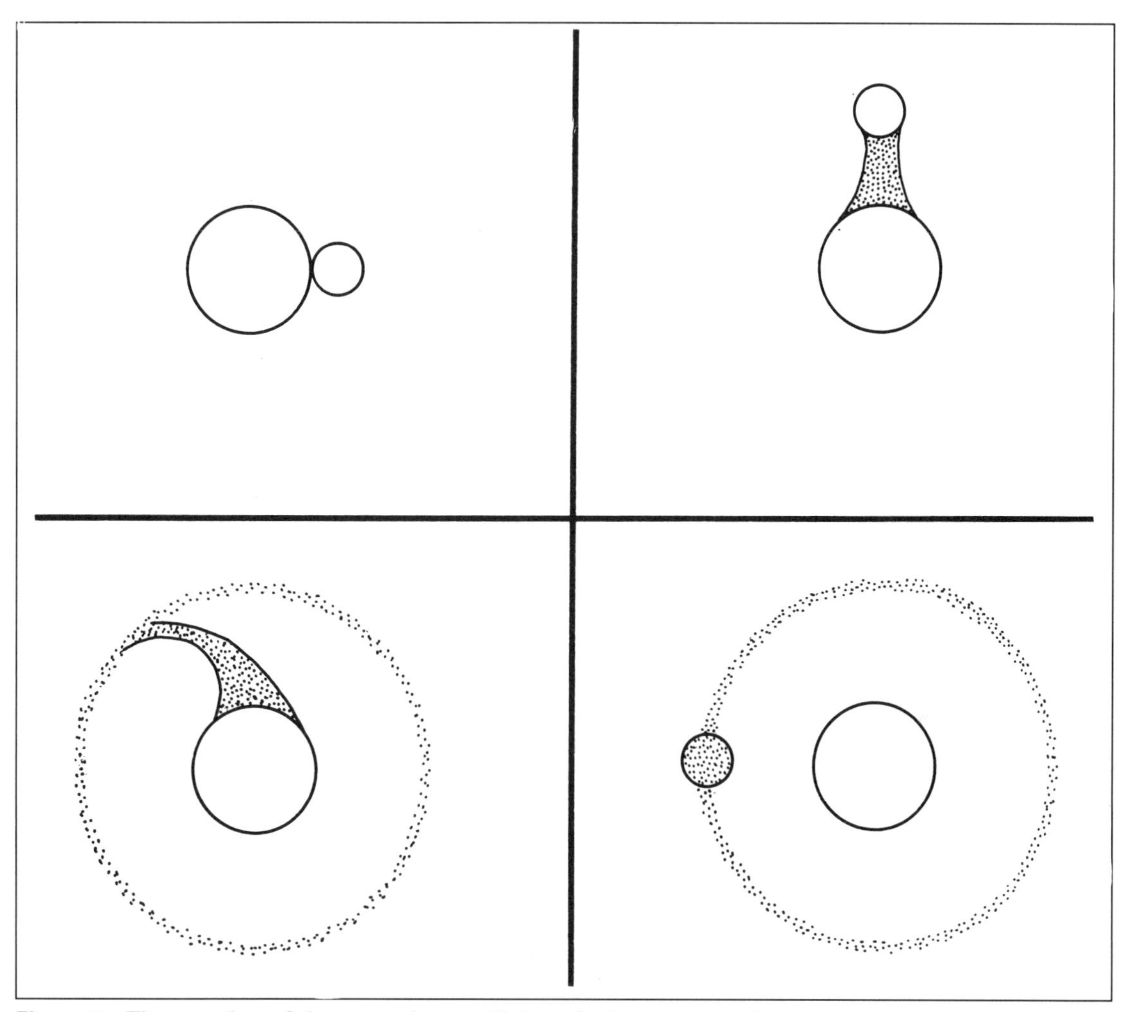

Figure 5 The creation of the moon by a collision of a large asteroid.

verted into large compact basins, with walls nearly 2 miles above the surrounding terrain and floors up to 10 miles deep.

The meteorites plunged into the planet's thin basaltic crust and gouged out huge quantities of partially solidified and molten rock. The scars in the crust quickly healed over as great batches of fresh magma bled through giant fissures and poured out onto the surface, creating a magma ocean. The massive floods of basalt were similar to those that formed the dark maria, or lava plains, on the moon soon after the large impacts subsided. When the Earth acquired an atmosphere and ocean, intense weather systems eroded all the remaining craters, and no telltale evidence of their existence remains today.

THE BIG SPLASH

While the Earth was forming out of bits of metal and rock, somewhere farther out in space was an asteroid about the size of Mars. Soon after the Earth's formation 4.6 billion years ago, the asteroid was knocked out of the asteroid belt between the orbits of Mars and Jupiter either by Jupiter's strong gravitational pull or by a collision with a wayward comet. On its way toward the inner Solar System, the asteroid glanced off the Earth (Figure 5). The impact resulted in a powerful explosion equivalent to the detonation of an amount of TNT equal to the mass of the asteroid.

The collision tore a great gash in the Earth, and a large portion of its molten interior along with some of the rocky mantle of the impactor spewed into orbit, forming a ring of debris around the planet called a protolunar disk. The force of the impact also might have knocked the Earth over, tilting its rotational axis 23 degrees. Similar collisions involving the other planets, especially Uranus, which lies on its side, might explain their various degrees of tilt and elliptical orbits. The glancing blow also might have sped up the Earth's rotation, and the increased angular momentum melted the planet throughout.

The Earth's new satellite continued to grow as it swept up the debris in orbit around the Earth. In addition, small satellites composed of rock fragments orbited the moon and crashed onto its surface. A massive shower of meteorites also bombarded the moon at the same time the Earth was heavily cratered, and numerous 50-mile-wide asteroids struck the lunar surface and broke through the thin crust. Great floods of dark basaltic lava spilled out onto the surface, giving the moon a landscape of giant craters and flat lava plains (Figure 6). Because of this heavy volcanic activity and the dark plains it produced, the moon reflects only 7 percent of its sunlight, making it one of the darkest bodies in the Solar System.

The moon from *Apollo 11* on July 1969. Courtesy of NASA

The moon became gravitationally locked onto the Earth, rotating on its own axis at the same rate it revolves around the Earth. Therefore, only one side faces its mother planet at all times. Many other moons in the Solar System share this characteristic, which suggests that they formed in a similar manner as the Earth's moon. The absence of a Venusian moon is very curious, however. It might have crashed into its mother planet or escaped into orbit around the sun. Perhaps Mercury, which is about the size of our moon, was once the moon of Venus.

The Earth's moon cooled quickly and formed a thick crust long before the planet did because its ratio of surface area to volume was much larger, allowing it to radiate more heat into space. Convection in the moon's mantle and molten iron core might have generated a weak magnetic field. However, convective motions were not strong enough to drive lithospheric plates as they do on Earth. During the great meteorite bombardment, the moon was highly cratered. But because it had no atmosphere or moving plates to erode the surface, it still retains most of its original terrain features.

Moon rocks brought back by astronauts during the Apollo missions of the late 1960s and early 1970s (Figure 7) are believed to have a similar composition as the Earth's upper mantle, and range in age from 3.2 to 4.5 billion years old. Because no rocks were found that dated younger than 3.2 billion years, the moon probably ceased volcanic activity at this time, and its interior began to cool and solidify.

Figure 7 A lunar sample brought back by Apollo astronauts, similar to ejected fragments found around impact craters. Photo by H. J. Moore, courtesy of USGS

After its formation, the moon closely orbited the Earth, filling much of the

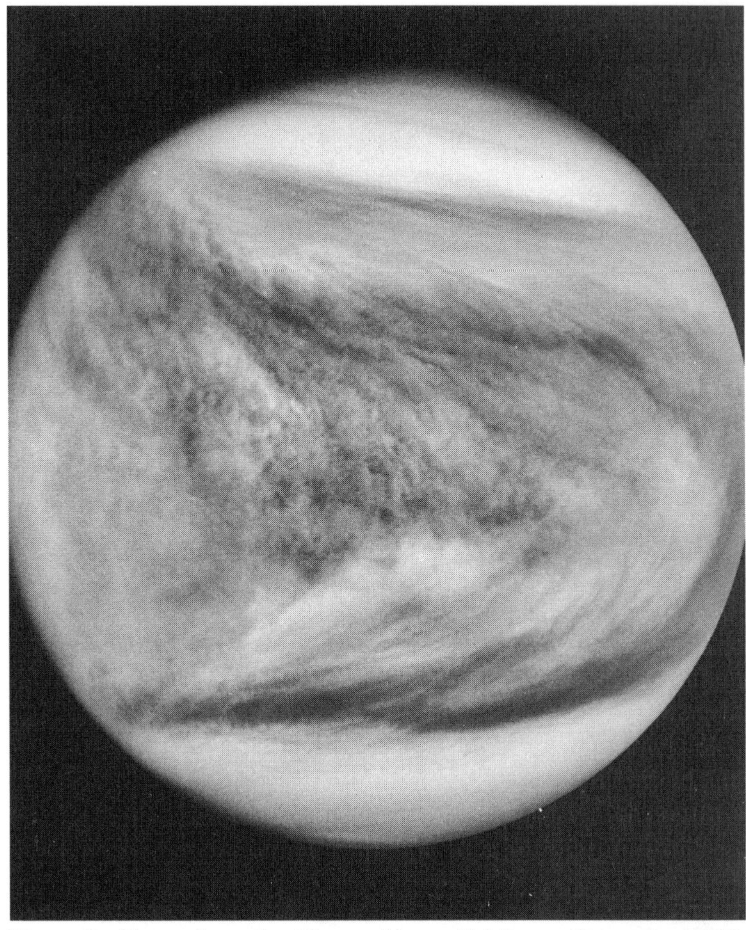

Figure 8 Venus from the *Pioneer Venus Orbiter* on December 1980.
Courtesy of NASA

sky, and its strong gravitational attraction caused huge tidal bulges in the Earth's thin crust. The Earth's rotation was more rapid than it is today, and days were only a few hours long. Over time, the moon was flung outward in an ever-widening orbit, slowing down the Earth's rotation. Even now, the moon is still moving away from the Earth at about an inch or so each year, and days continue to get a little longer.

THE BIG BURP

Early in the formation of the Earth, during a period geologists like to refer to as the "big burp," giant volcanoes belched out massive quantities of gas and steam. In addition, icy projectiles from outer space plunged into the planet, releasing huge volumes of gas and water vapor, and by some accounts this was the main source of the atmosphere and ocean. The result was a thick, steamy atmosphere, with atmospheric pressures 100 times greater than they are today. Massive clouds covered the entire planet as they presently do on Venus (Figure 8). Indeed, Venus with its heavy carbon dioxide atmosphere is used as a model for the Earth's early years.

Some 4 billion years ago, as the Earth was cooling, heavy rains fell out of the clouds like massive sponges being wrung dry. The torrential rains produced the greatest floods the planet has ever known. Deep meteorite craters rapidly filled like huge bowls of water that spilled over onto flat lava plains. Giant canyons were carved out as water rushed down the steep sides of colossal volcanoes, which continued to spew steam and gases into the atmosphere. In addition, multitudes of icy comets added their own ingredients to the deluge. When the skies finally cleared, the Earth was transformed into a glistening blue orb, covered almost entirely by a deep global ocean.

As yet, no continents marred the Earth's watery face, and only a small number of volcanoes rose high enough to dot the seascape with a few scattered islands. An alien world existed on the floor of the ocean, where volcanoes continued to erupt undersea. Hydrothermal vents dislodged black or milky colored water containing sulfur and other chemicals. In just a short time, by geologic standards, the sea turned from fresh to salty and contained all the necessary ingredients for the emergence of life.

Life arose on this planet during a period of crustal formation and outgassing (the loss of gas stored within a planet) of an atmosphere and ocean. This was also a time of heavy meteorite bombardment, which might have influenced the final outcome of the planet. When proteins were first striving to organize into living cells, they found conditions extremely difficult because the Earth was constantly being showered by comets and meteorites. The early cells might have been repeatedly exterminated, forcing life to originate over and over again. The larger impactors might have boiled away the entire ocean, creating surface temperatures of over 200 degrees Fahrenheit and pressures of over 100 atmospheres, thus turning the planet into a pressure cooker that exterminated all life.

THE ASTEROID BELT

Between the orbits of Mars and Jupiter lies the asteroid belt, comprising about a million pieces of Solar System rubble larger than a mile across as

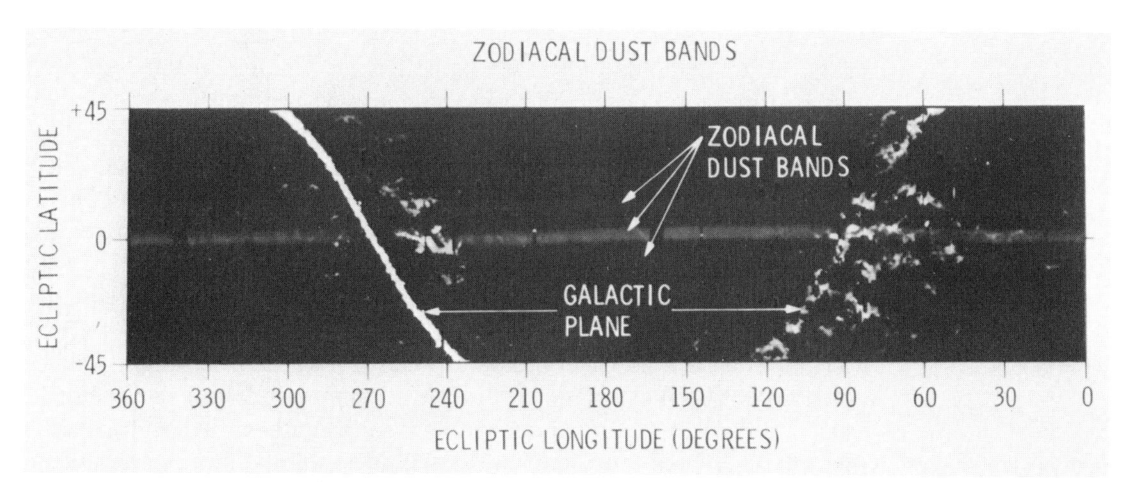

Figure 9 The zodiacal dust bands are debris from comets and collisions between asteroids near the inner edge of the asteroid belt. Courtesy of NASA

TABLE 1 SUMMARY OF MAJOR ASTEROIDS

Asteroid	Diameter (miles)	Distance from Sun (million miles)	Type
Ceres	635	260	Carbon-rich
Pallas	360	258	Rocky
Vesta	344	220	Rocky
Hygeia	275	292	Carbon-rich
Interamnia	210	285	Rocky
Davida	208	296	Carbon-rich
Chiron	198	1270	Carbon-rich
Hektor	130	480	Uncertain
Diomedes	118	472	Carbon-rich

well as more numerous smaller objects. Zodiacal dust bands consisting of fine material orbit the sun near the inner edge of the main asteroid belt (Figure 9). The debris is believed to have originated from comet trails, composed of dust and gas blown outward by the solar wind, and from collisions between asteroids.

Asteroids are leftovers from the creation of the Solar System, and due to the strong gravitational attraction of Jupiter they were unable to coalesce into a single planet. Instead, they formed several planetoids smaller than the moon, as well as a broad band of debris, called meteoroids, which are small fragments broken off of asteroids by numerous collisions. Originally, the combined masses of all the material in the asteroid belt was nearly equal to the present mass of the Earth. However, constant collisions have weeded out the asteroids so that now their combined mass is less than 1 percent of the original.

Some of the large asteroids might have melted and differentiated early in the formation of the Solar System. Inner and middle belt asteroids underwent a great deal of heating and experienced as much melting as did the planets. The molten metal in the asteroids along with siderophiles (iron lovers), such as iridium and osmium of the platinum group, sank to their interiors and solidified. The metallic cores were exposed after eons of collisions between asteroids chipped away the more fragile surface rock. Thus, breakup after collisions yielded several dense, solid fragments (Figure 10). A large portion of the asteroids contain a high concentration of iron and nickel, suggesting they were once part of the metallic core of a planetoid that disintegrated after a collision with another body.

Figure 10 A planetoid smaller than the moon is broken up by a collision with another body, and additional collisions yield asteroids that bombard the Earth.

The stony asteroids, which are much less dense and contain a high percentage of silica, exist near the inner part of the asteroid belt. The darker carbonaceous asteroids, which contain a high percentage of carbon, lie toward the outer part of the asteroid belt. In between these regions are wide spaces called Kirkwood gaps, named for American mathematician Daniel Kirkwood, that are almost totally devoid of asteroids. If an asteroid falls into one of these gaps, its orbit stretches, causing it to swing in and out of the asteroid belt, which brings it close to the sun and the orbits of the inner planets.

ASTEROIDS AND COMETS

Asteroids and comets are distinctly different inhabitants of the Solar System. Surrounding the sun (about a light-year away) is a shell of over 1 trillion comets with a combined mass of 25 Earths, called the Oort Cloud, after Dutch astronomer Jan Kendrick Oort. Another band of comets might exist closer to the sun but still well beyond Pluto, which due to its odd orbit might itself be a captured comet nucleus or an asteroid.

Figure 11 Comet Halley from the National Optical Astronomy Observatories on April 1986. Courtesy of NOAO

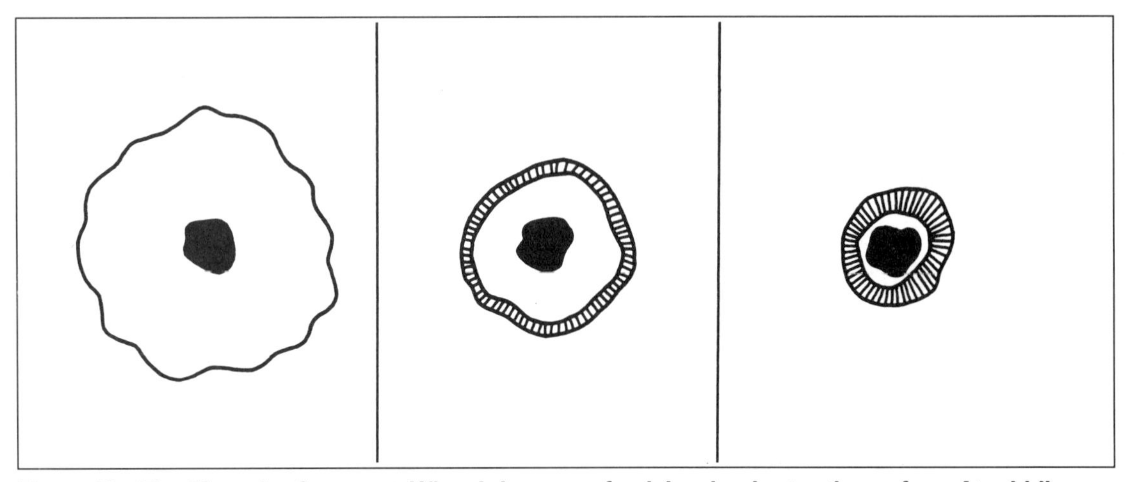

Figure 12 The life cycle of a comet: When it is young, fresh ice dominates the surface. At middle age, the comet develops an insulated crust. During old age, the crust becomes thick enough to cut off all cometary activity.

Comets, like Comet Halley (Figure 11), which appeared in 1985–1986 after spending 76 years in deep space, are hybrid planetary bodies consisting of a stony inner core and an icy outer layer. They are often described as dirty snowballs composed of ice mixed with rock fragments. Some comets have highly elliptical orbits that take them within the inner Solar System. As they approach the inner Solar System, water vapor and gases stream outward, forming a tail millions of miles long that points away from the sun due to the outflowing solar wind.

Apollo and Amor asteroids are Earth-crossing asteroids that possibly began their lives as comets. Through eons, their coating of ices and gases has been eroded by the sun, exposing what appear to be large chunks of rock (Figure 12). They are not confined to the asteroid belt as are the great majority of known asteroids but instead approach or even cross the orbit of the Earth. Usually within a few tens of millions of years, the Apollos either collide with one of the inner planets or are flung out in wide orbits after a near miss.

Dozens of Apollo asteroids have been identified out of a possible total of perhaps 1,000. Most are quite small and discovered only when they swing close by the Earth (Table 2). Many of these Earth-crossing asteroids do not originate in the asteroid belt but are believed to be comets that have exhausted their volatile material after repeated encounters with the sun and have lost their ability to produce a coma or tail. Inevitable collisions with the Earth and the other inner planets are steadily depleting them, requiring an ongoing source of new Apollo-type asteroids either from the asteroid belt or from the contribution of burned-out comets.

TABLE 2 CLOSEST CALLS WITH EARTH

Body	Distance in A.U. (Earth–Moon)	Date
1989 FC	0.0046 (1.8)	March 22, 1989
Hermes	0.005 (1.9)	October 30, 1937
Hathor	0.008 (3.1)	October 21, 1976
1988 TA	0.009 (3.5)	September 29, 1988
Comet 1491 II	0.009 (3.5)	February 20, 1491
Lexell	0.015 (5.8)	July 1, 1770
Adonis	0.015 (5.8)	February 7, 1936
1982 DB	0.028 (10.8)	January 23, 1982
1986 JK	0.028 (10.8)	May 28, 1986
Araki-Alcock	0.031 (12.1)	May 11, 1983
Dionysius	0.031 (12.1)	June 19, 1984
Orpheus	0.032 (12.4)	April 13, 1982
Aristaeus	0.032 (12.4)	April 1, 1977
Halley	0.033 (12.8)	April 10, 1837

For a comet to evolve into an asteroid, it must enter into a stable orbit in the inner part of the Solar System. Meanwhile, its activity is reduced so that it becomes a burned-out hulk composed mostly of rock. Even when a comet encounters one of the large outer planets like Jupiter and is trapped in a short-period orbit, its path around the sun is rarely stable. Soon it reencounters the giant planet and is flung back into deep space, possibly escaping the Solar System altogether.

After a comet achieves a stable short-period orbit, it makes repeated passages close to the sun. Every time it passes by the sun, it loses a few feet of its outer layers. The solar wind forces gas and dust particles to stream away from the comet, but the heavier silicate particles are pulled back into the nucleus by the comet's weak gravity. Gradually, an insulating crust forms to protect the icy inner regions of the nucleus from the sun's heat, and the comet ceases its outgassing and masquerades as an asteroid.

HISTORICAL IMPACTS

Around 4 billion years ago, a giant meteorite impact supposedly triggered the evolution of ancient continental shields, upon which the continents grew. During the height of the great meteorite bombardment, a massive asteroid landed in what is now central Ontario, Canada (Figure 13). The impact might have created a crater up to 900 miles wide and triggered the formation of continental masses, which previously had been missing on the Earth's surface.

In Canada, Australia and Africa, rocks that were exposed on the surface during the first 2 billion years have remained relatively unchanged. Layers of tiny spherical grains called spherules are suspected of being debris from the oldest known meteorite impacts, occurring around 3.5 billion years ago. In South Africa's Barberton greenstone belt lies a thick, widespread bed of silicate spherules, which are believed to have originated from the melt of a large meteorite impact sometime between 3.5 and 3.2 billion years ago. Another huge impact in South Africa formed the wide Vredefort impact structure. The melted materials from the impact had a high concentration of the rare element iridium, an isotope of platinum, indicating that the crater had an extraterrestrial origin.

About 1.8 billion years ago, a large meteorite slammed into the North American continent in Ontario, Canada, and generated enough energy to melt vast quantities of basalt and granite. Metals separating out of the molten rocks formed the world's largest and richest nickel ore deposit, known as the Sudbury Igneous Complex. This location is also one of the world's oldest astroblemes, which are ancient eroded impact structures. The main line of evidence that the Sudbury complex was formed by a meteorite impact is shatter cones, which are distinctively striated conical rocks fractured by shock waves and found only at known meteorite impact structures.

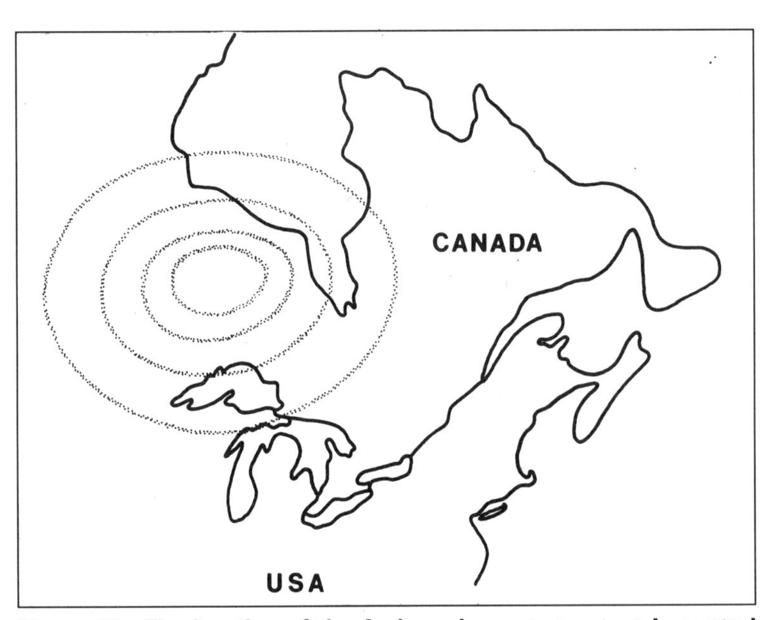

Figure 13 The location of the Archean impact structure in central Ontario, Canada.

A 30-mile-wide rimmed circular structure lies more than a mile beneath the floor of Lake Huron. It appears to be an impact crater that was blasted out by a large meteorite at least 500 million years ago. The ring structure was detected with magnetic sensors and named the Can-Am structure because it straddles the Canadian and U.S. border. Such a crater would have required the impact of a meteorite about 3 miles in diameter and would number among at least 100 other large craters from impacts during the last 500 million years.

At the beginning of the Triassic period, around 210 million years ago, a huge asteroid rammed into the Earth, creating the 60-mile-wide Manicouagan impact structure in Quebec, Canada (Figure 14). The impact has been blamed for a mass extinction that killed nearly half of the reptile families. This paved the way for the evolution of the direct ancestors of modern animals except for the birds, which did not arrive for another 50 million years. In addition, the dinosaurs emerged to become masters of the Earth for the next 145 million years (Table 3).

Around 65 million years ago, another asteroid is thought to have struck the Earth, creating an explosion thousands of times stronger than all the nuclear weapons in the world and tearing a hole in the crust at least 100 miles wide. An incredible amount of debris was lofted high into the atmosphere, shutting out the sun and placing the Earth in a deep freeze. The dinosaurs probably could not have tolerated these sudden frigid conditions. Instead of a single large impact, a swarm of meteorites or comets might have peppered the Earth. Five craters have thus far been identified that date to the end of the dinosaur era.

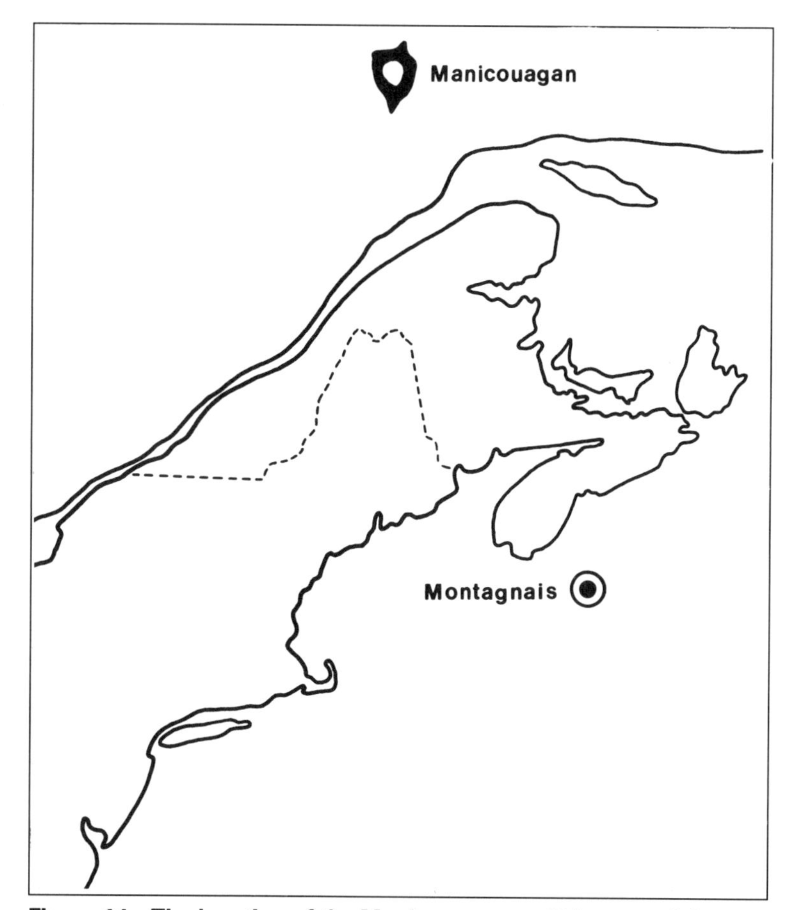

Figure 14 The location of the Manicouagan and Montagnais impact structures in North America.

TABLE 3 THE GEOLOGIC TIME SCALE

Era	Period	Epoch	Age (millions of years)	First Life Forms
	Quaternary	Holocene	0.01	
		Pleistocene	2	Man
Cenozoic		Pliocene	7	Mastodons
		Miocene	26	Saber-tooth tigers
	Tertiary	Oligocene	37	
		Eocene	54	Whales
		Paleocene	65	Horses Alligators
	Cretaceous		135	Birds
Mesozoic	Jurassic		190	Mammals
	Triassic		240	Dinosaurs
	Permian		280	Reptiles
		Pennsylvanian*	310	
	Carboniferous			Trees
Paleozoic		Mississippian	345	Amphibians Insects
	Devonian		400	Sharks
	Silurian		435	Land Plants
	Ordovician		500	Fish
	Cambrian		570	Sea Plants Shelled animals
			700	Invertebrates
Proterozoic			2500	Metazoans
			3500	Earliest life
Archean			4000	Oldest rocks
			4600	Meteorites

*Note: Pennsylvanian and Mississippian represent geologic changes that occurred in North America over the Carboniferous.

If the asteroid had landed in the sea, it is conceivable that it disappeared along with the oceanic crust upon which it impacted due to plate tectonics, which is the interaction of movable crustal plates on the Earth's surface. This process removes the ocean floor by plate subduction, which is the thrusting of the oceanic crust into the Earth's interior, called the mantle. Since the end of the Cretaceous, roughly half of the oceanic crust has been recycled into the mantle. Nevertheless, what appears to be an impact site that is mostly intact is located 300 miles northeast of Madagascar off the east coast of Africa. The impact site also has the correct age, around 65 million years old, to make it a prime candidate for the dinosaur killer.

An asteroid 10 miles or more wide would be required to have gouged out the 200-mile-wide circular depression. The impact might have also triggered the Deccan Traps volcanism in India, which was in the vicinity at the time and heading for Asia (Figure 15). The Deccan Traps are the greatest outpouring of basalt on land over the last 250 million years. The massive eruptions that created them might have had a hand in the demise of the dinosaurs by drastically changing the environment.

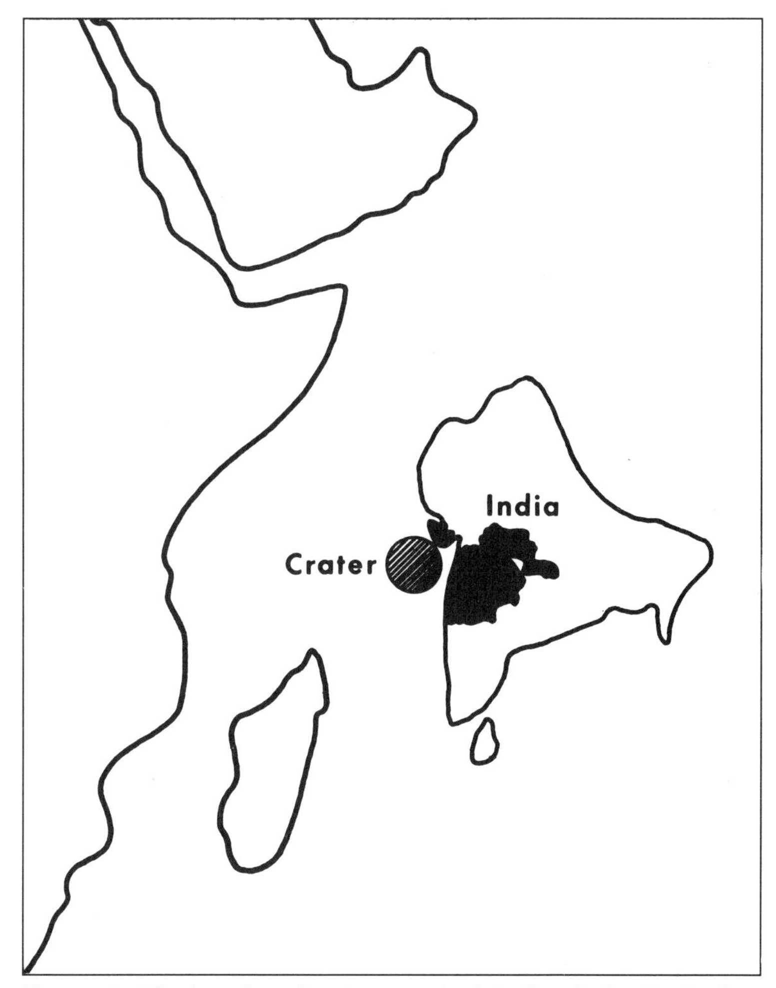

Figure 15 The location of a crater south of the Seychelles Bank when India was drifting towards Asia 65 million years ago.

Around 40 million years ago, two, or perhaps three, large meteorite impacts might have caused the extinction of the archaic mammals. One of these impacts might have created the Everglades on the southern tip of Florida. The Everglades is a swamp and forested area surrounded by an oval-shaped system of ridges. A thick layer of limestone in the surrounding areas, around 40 million years old, is suspiciously missing over most of the southern part of the Everglades. Apparently, a large meteorite slammed into the limestones, which were submersed under 600 feet of water, and fractured the rocks. The impact also would have generated an enormous tsunami, or sea wave, that swept the debris far out to sea.

During the height of the last ice age about 22,000 years ago, a large meteorite impacted in northern Arizona and vaporized, forming a large mushroom-shaped cloud of debris. The meteorite ejected nearly 200 million tons of rock and excavated a crater measuring 4,000 feet across and nearly 600 feet deep (Figure 16). The crater rim, formed by upraised sedimentary layers, rises to 150 feet above the desert floor. Pulverized rock blanketed the area around the crater to a depth of up to 75 feet.

The largest known meteorite crater, where actual meteoritic debris has been found, is the New Quebec Crater in Quebec, Canada. It has a diameter of about 11,000 feet and is about 1,300 feet deep. The crater presently contains a deep lake, the surface of which is 500 feet below the crater rim. The crater is estimated to be only a few thousand years old. It is well preserved because few changes occur in the frigid tundra.

A relatively young crater is located on the northern edge of the Great Sandy Desert in Western Australia south of Halls Creek. The Wolf Creek

Figure 16 Meteor Crater, Coconino County, Arizona. Photo by W. B. Hamilton, courtesy of USGS

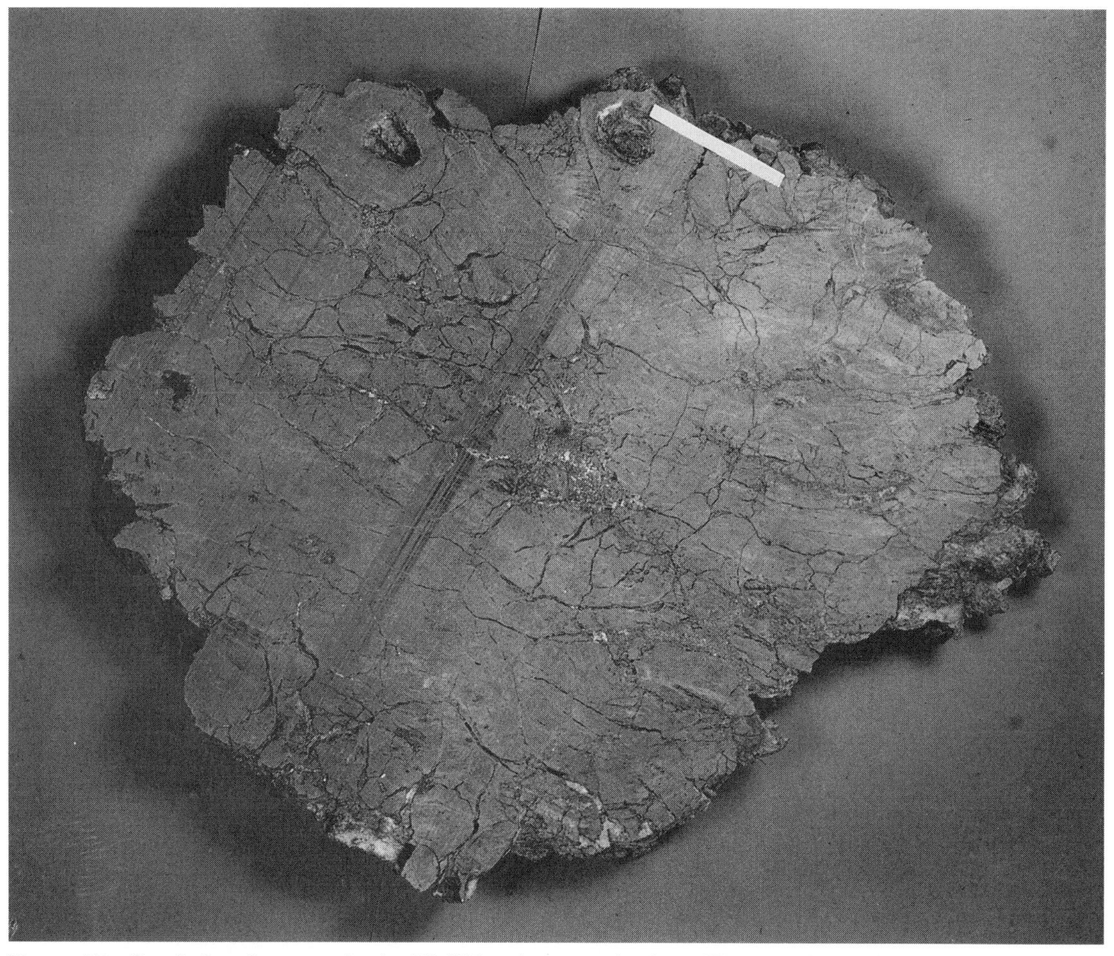

Figure 17 Crack development in the Wolf Creek meteorite from Western Australia. Photo by G. T. Faust, courtesy of USGS

Crater has a diameter of 2,800 feet and a depth of 140 feet. Several large pieces of a meteorite, some weighing over 300 pounds, were discovered near the crater (Figure 17). Because the crater is located in the desert where little erosion occurs, it also is well preserved. However, many similar young craters might be scattered around the world that were not so lucky. This suggests that this type of cratering is an ongoing process and that the Earth can expect another major impact some time in the future.

2

IMPACT CRATERS

It is difficult searching land and sea for ancient impact craters because the Earth's highly active geology has long since erased all but their faintest signs. Impacts on the Earth's moon, the inner planets, and the moons of the outer planets are quite evident and numerous. The Earth was probably hit by more than a dozen times as many meteorites as the moon because of the planet's larger size and greater gravitational attraction. However, the moon retains a better record of terrestrial impact cratering than the Earth itself. But many of the lunar craters tend to overlap each other, which erases any regular pattern and makes it difficult to date them accurately.

Fortunately, several remnants of ancient craters still remain on Earth, which suggests that it was just as heavily bombarded as the rest of the Solar System. Many strikingly circular features have been found that appear to be impact craters. Due to their low profiles and subtle stratigraphy they were previously unrecognized as impact structures. In the future, many more craters are bound to be discovered through the use of sophisticated instruments aboard satellites, to provide a clearer picture of what transpired long ago.

CRATERING RATES

One of the purposes of planetary science is to compare the geologic histories of the planets and their moons by establishing a relative time scale based on the record of impact cratering. Generally, the older the surface, the more craters that are on it. The heavily cratered lunar highland is the most ancient region on the moon. It contains a record of intense bombardment around 4 billion years ago. Since then, the number of impacts rapidly declined, and the impact rate has remained relatively low. If the impacts had continued at a high rate throughout the Earth's history, the evolution of life would have been substantially altered.

The rate of cratering also appears to differ from one part of the Solar System to another. The cratering rates by asteroids and comets along with the total number of craters suggests that the average rates over the past few billion years were similar for the Earth, its moon, and the rest of the inner planets. However, the cratering rates for the moons of the outer planets might have been substantially lower than the cratering rates for the inner Solar System. Nevertheless, the size of the craters in the outer Solar System are comparable to those in the inner Solar System (Figure 18).

Figure 18 The heavily cratered terrain on Mercury from *Mariner 10* in March 1974. Courtesy of NASA

The cratering rates for the moon and Mars were nearly the same, except that on Mars erosional agents such as wind and ice might have erased many of its craters (Figure 19). Indeed, what appear to be ancient stream channels flowing away from craters seem to indicate that at some time Mars had running water. On the moon, however, the dominant mechanism for destroying craters is other impacts. There is so much crater overlap, it is often difficult to place the craters in their proper geologic order. The impact rates for Mars might have actually been higher than those for the moon, possibly because Mars is much closer to the asteroid belt. Major obliteration events have occurred on Mars as recently as 200 to 450 million years ago, whereas most of the scarred lunar terrain was produced billions of years ago.

Figure 19 A heavily cratered region on Mars showing the effects of wind erosion, from *Viking Orbiter 1* in June 1980. Courtesy of NASA

Impact craters on Earth range in age from a few thousand to nearly 2 billion years old. For the past 3 billion years, the cratering rate for the Earth has been fairly constant, with a major impact, resulting in a crater 30 miles or more in diameter, occurring every 50 to 100 million years. As many as three large meteorite impacts, producing craters with diameters of at least 10 miles, are expected every million years. Major meteorite impacts also appear to be somewhat periodic, occurring every 26 to 32 million years. This might also account for the extinction of species on a similar time scale.

TABLE 4 LOCATION OF MAJOR METEORITE IMPACT STRUCTURES

Name	Location	Diameter (feet)
Al Umchaimin	Iraq	10,500
Amak	Aleutian Islands	200
Amguid	Sahara Desert	
Aouelloul	Western Sahara Desert	825
Bagdad	Iraq	650
Boxhole	Central Australia	500
Brent	Ontario, Canada	12,000
Campo del Cielo	Argentina	200
Chubb	Ungava, Canada	11,000
Crooked Creek	Missouri, USA	
Dalgaranga	Western Australia	250
Deep Bay	Saskatchewan, Canada	45,000
Dzioua	Sahara Desert	
Duckwater	Nevada, USA	250
Flynn Creek	Tennessee, USA	10,000
Gulf of St. Lawrence	Canada	
Hagensfjord	Greenland	
Haviland	Kansas, USA	60
Henbury	Central Australia	650
Holleford	Ontario, Canada	8,000
Kaalijarv	Estonia	300
Kentland Dome	Indiana, USA	3,000
Kofels	Austria	13,000
Lake Bosumtwi	Ghana	33,000

Name	Location	Diameter (feet)
Manicouagan Reservoir	Quebec, Canada	200,000
Merewether	Labrador, Canada	500
Meteor Crater	Arizona, USA	4,000
Montagne Noire	France	
Mount Doreen	Central Australia	2,000
Murgab	Tadjikstan	250
New Quebec	Quebec, Canada	11,000
Nordlinger Ries	Germany	82,500
Odsessa	Texas, USA	500
Pretoria Saltpan	South Africa	3,000
Serpent Mound	Ohio, USA	21,000
Sierra Madera	Texas, USA	6,500
Sikhote-Alin	Siberia	100
Steinheim	Germany	8,250
Talemzane	Algeria	6,000
Tenoumer	Western Sahara Desert	6,000
Vredefort	South Africa	130,000
Wells Creek	Tennessee, USA	16,000

Over 120 known impact craters are scattered around the world, most of which are younger than 200 million years (Table 4). Even though the cratering rate has been rather constant during the past 3 billion years, older craters are less abundant because they were destroyed by erosion or sedimentation. Thus far, only 10 percent of the large craters that are younger than 100 million years have been discovered. About two thirds of the known impact craters are located in stable regions known as cratons, composed of strong rocks in the interiors of continents (Figure 20). The cratons experience low rates of erosion and other destructive processes, allowing craters to be preserved for long periods.

METEORITE IMPACTS

The most accepted theory for the origin of meteorites is that they come from a jumble of asteroids in a wide belt lying between the orbits of Mars and Jupiter. Some asteroids might be the rocky cores of dead comets that lost their coating of ice by evaporation and settled into orbit around the sun. A few rare meteorites found on the ice sheets of Antarctica might be pieces of the Martian crust blasted out by large asteroid impacts. Even pieces of the moon might have landed on the Earth when major asteroid impacts blasted them into space.

Asteroids range in size from about 1 mile to several hundred miles in diameter (Figure 21). These large chunks along with numerous smaller fragments in the asteroid belt are believed to have a total mass of less than 1 percent of the Earth's mass. Even a 1-mile-wide asteroid could do significant damage upon striking the Earth's surface. But it remains a mystery how these large rock fragments managed to get into orbits that cross our planet's path. The asteroids seem

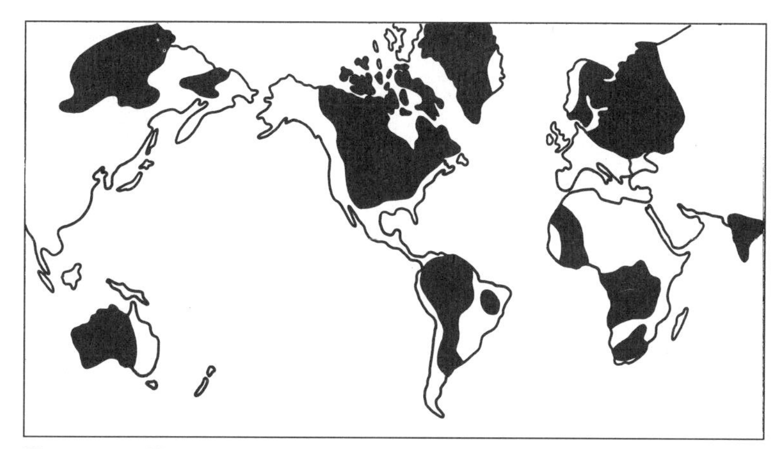

Figure 20 The worldwide distribution of stable cratons.

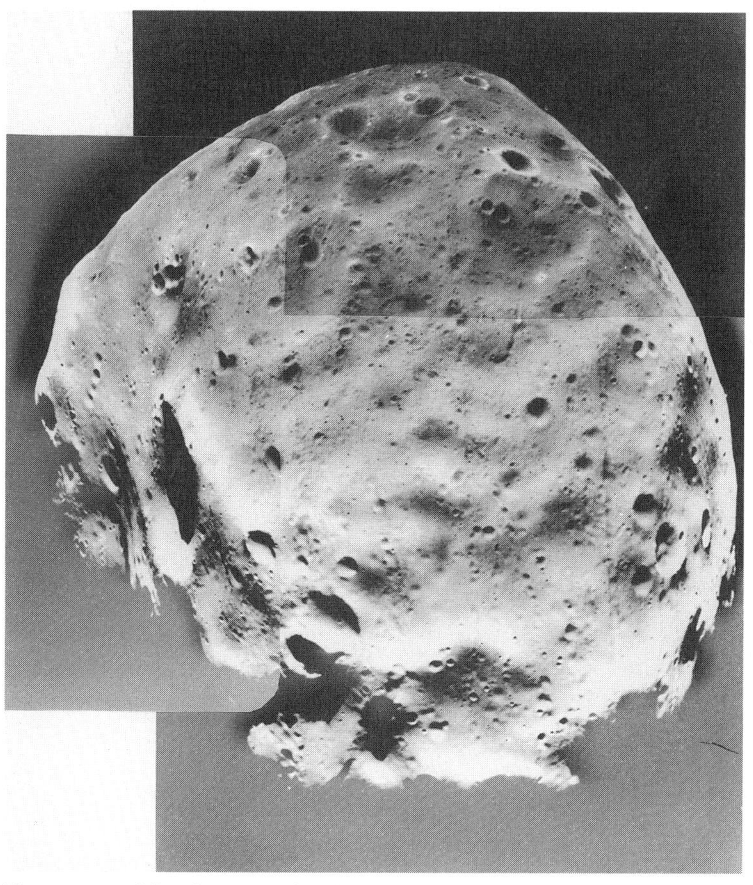

Figure 21 Mars' moon Phobos, measuring 13 miles across, is thought to be a captured asteroid. Courtesy of NASA

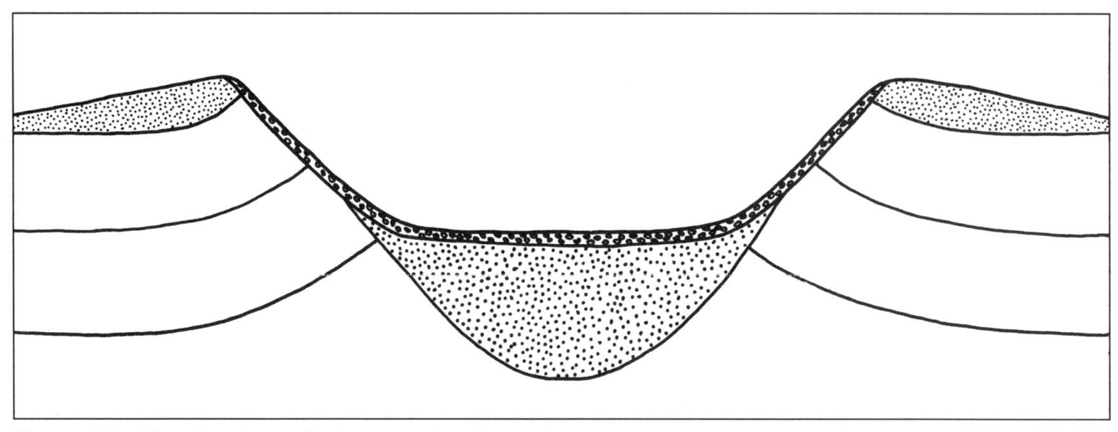

Figure 22 The structure of a large meteorite crater.

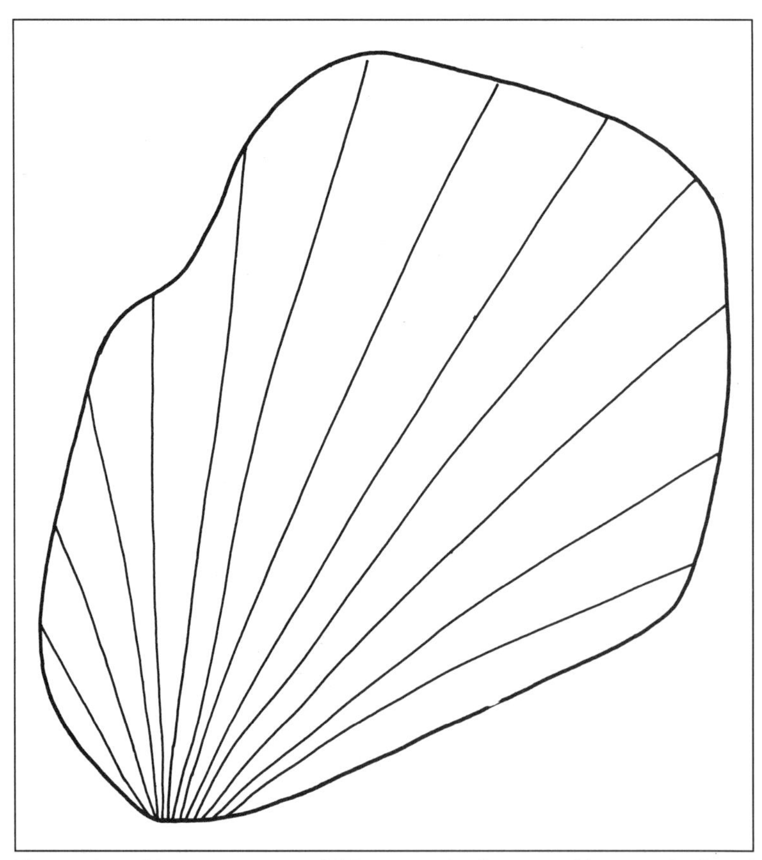

Figure 23 Shatter cones, which are rocks fractured in a conical and striated pattern in fine-grained rocks, are the most apparent effect of a large meteorite impact.

to be fairly stable and run in nearly circular orbits for millions of years. Then for unknown reasons, their orbits stretch out and become so elliptical that some actually come close to the Earth.

If a large asteroid slammed into the planet, it would expel a huge amount of sediment and produce a deep crater (Figure 22). The finer material would be lofted high into the atmosphere, while the coarse debris would fall back around the perimeter of the crater, forming a high, steep-banked rim. Not only would rocks be shattered in the vicinity of the impact, but the shock wave passing through the ground would produce shock metamorphism in the surrounding rocks, changing their composition and crystal structure.

The most easily recognizable shock effect is the fracturing of rocks into conical and striated patterns called shatter cones (Figure 23). They form most readily in fine-grained rocks that have little internal structure, such as limestone and quartzite.

The high temperatures developed by the force of the impact also fuse sediment into small glassy spherules, which are tiny spherical bodies. Extensive deposits, over 1 foot thick, of 3.5-billion-year-old spherules exist in South Africa. Spherules of a similar age also have been found in Western Australia. The spherules resemble the glassy chondrules (rounded granules) in carbonaceous chondrites, which are carbon-rich meteorites, and in lunar soils. The discoveries suggest that massive meteorite bombardments during the early part of the Earth's history played a major role in shaping the surface of the planet. In addition, some carbon-rich meteorites might have provided the necessary ingredients for the initiation of life.

Large meteorite impacts also produce shocked quartz grains with prominent striations across crystal faces (Figure 24). Minerals such as quartz and feldspar develop these features when high-pressure shock waves exert shearing forces on their crystals, producing parallel fracture planes called lamellae.

Sediments dating 65 million years old located at the boundary between the Cretaceous and Tertiary periods throughout the world (Figure 25) mark the extinction of the dinosaurs and many other species. They contain shocked quartz grains with distinctive lamellae, common soot from global forest fires set ablaze by glowing bits of impact debris flying past and back through the atmosphere, and unique concentrations of iridium, which is relatively abundant on meteorites and comets but practically nonexistent in the Earth's crust. Two rare amino acids known to exist only on meteorites were also found in the sediment layer. In addition, the sediments contained the mineral stishovite, a dense form of silica found nowhere on Earth except at known impact sites.

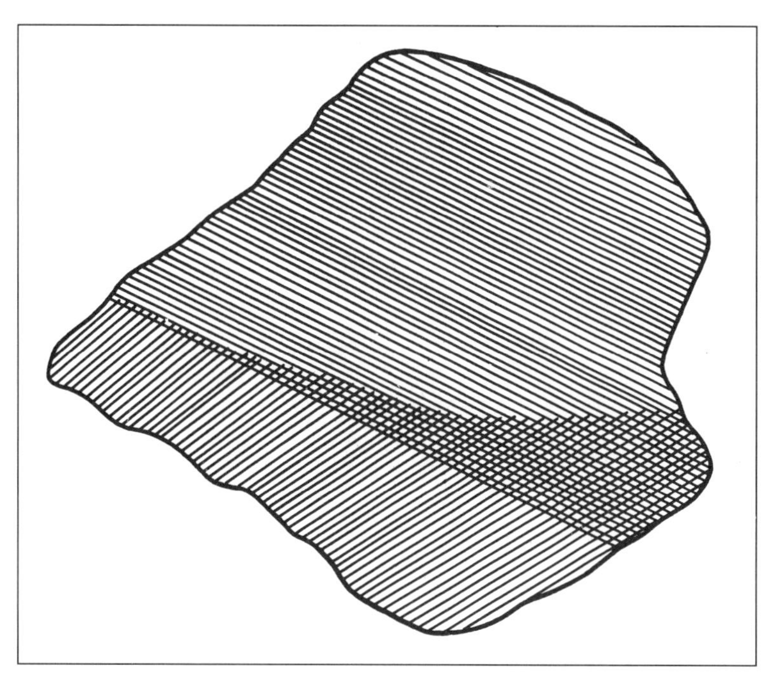

Figure 24 Lamellae are striations across crystal faces produced by high pressure shock waves from a large meteorite impact.

Figure 25 The southwest slope of South Table Mountain, Golden, Colorado. The boundary between the Cretaceous and Tertiary periods lies 10 feet below where the man is standing. Photo by R. W. Brown, courtesy of USGS

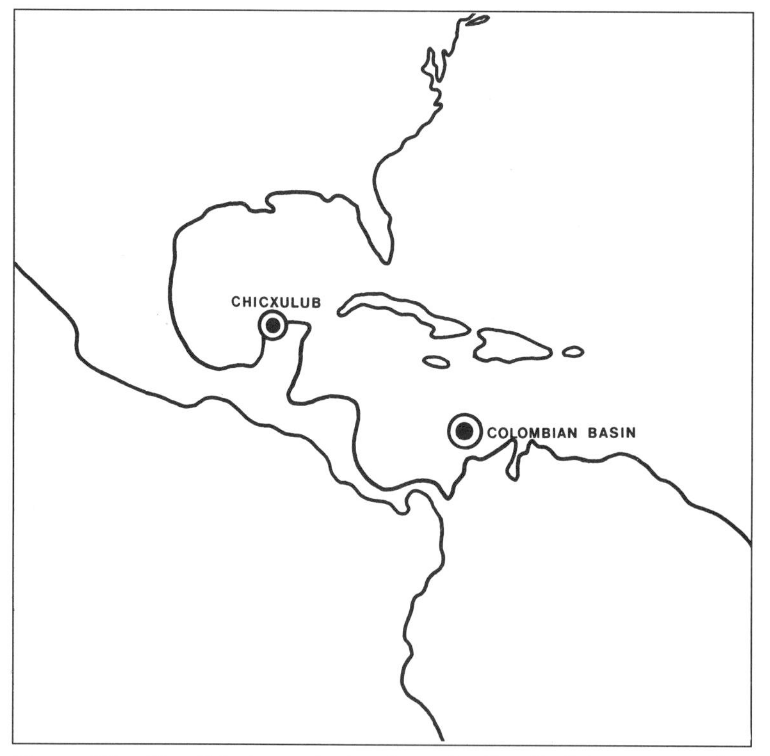

Figure 26 Possible impact structures in the Caribbean area that might have ended the Cretaceous period.

Much of the search for the meteorite impact site has been concentrated around the Caribbean area (Figure 26), where thick deposits of wave-deposited rubble have been found along with melted and crushed rock ejected from the crater. If the meteorite landed on the seabed just off shore, 65 million years of sedimentation would have long since buried it under thick deposits of sand and mud. Furthermore, a splashdown in the ocean would have created an enormous sea wave, or tsunami, that would scour the seafloor and deposit its rubble on nearby shores.

CRATER FORMATION

Meteor Crater, also known as Barringer Crater, located 15 miles west of Winslow, Arizona, is one of the most spectacular impact craters on Earth (Figure 27). It is 4,000 feet across and 560 feet deep and was first mistaken for a volcanic crater because volcanoes were once prevalent in the region. However, its distinctive appearance more closely resembles craters on the moon. Boreholes drilled in

Figure 27 Meteor Crater, Coconino County, Arizona. Courtesy of USGS

the center of the crater and on the south rim, rising 135 feet above the desert floor, failed to locate the meteorite. But scattered in all directions outward from the crater were several tons of metallic meteoritic debris, indicating that the meteorite was of the iron-nickel variety, measuring about 200 feet in diameter and weighing about 1 million tons. The impactor released the equivalent energy of about 20 megatons of TNT, equal to the most powerful nuclear weapons.

Large meteorites traveling at high velocities completely disintegrate upon impact. In the process, they create craters that are generally 20 times wider than the meteorites themselves. A large meteorite impact sends out a shock wave with pressures of millions of atmospheres down into the rock and back up into the meteorite. As the meteorite burrows into the ground, it forces the rock aside and flattens itself in the process. It is then deflected and its shattered remains are thrown out of the crater, along with a spray

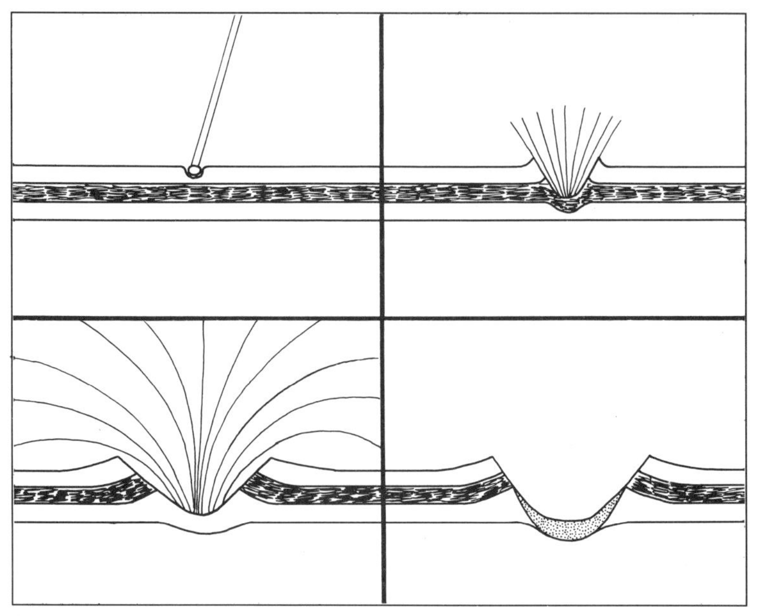

Figure 28 The formation of a large meteorite crater.

of shock-melted meteorite and melted and vaporized rock that shoots out at a high velocity (Figure 28).

As the spray continues to rise, it forms a rapidly expanding plume. The plume grows to several thousand feet across at the base, while the top extends several miles into the atmosphere. Most of the surrounding atmosphere is blown away by the tremendous shock wave created by the meteorite impact. The giant plume turns into an enormous black dust cloud that punches its way through the atmosphere like the mushroom cloud from a hydrogen bomb explosion. In fact, there are striking similarities between the effects of nuclear detonations and large meteorite impacts.

The crater diameter varies with the type of rock that is impacted due to the relative differences in rock strength. A crater made in crystalline rock can be twice as large as one made in sedimentary rock. Simple craters such as Meteor Crater form deep basins and range up to 2.5 miles in diameter. Larger craters, called complex craters, are much shallower and up to 100 times wider than they are deep. They generally have an uplifted structure in the center surrounded by an annular trough and a fractured rim similar to the central peaks inside craters on the moon as well as those on the seafloor.

IMPACT STRUCTURES

Numerous impact structures are scattered around the world (Figure 29). They are large circular features created by the sudden shock of a massive meteorite landing on the surface and are about 1 to 50 miles or more wide. Some impacts form distinctive craters, while others might show only subtle outlines of former craters. In this case, the only evidence of their existence might be a circular disturbed area, where the rocks were altered by shock metamorphism. Shock metamorphism requires the instantaneous application of high temperatures and pressures like those found deep in the Earth's interior.

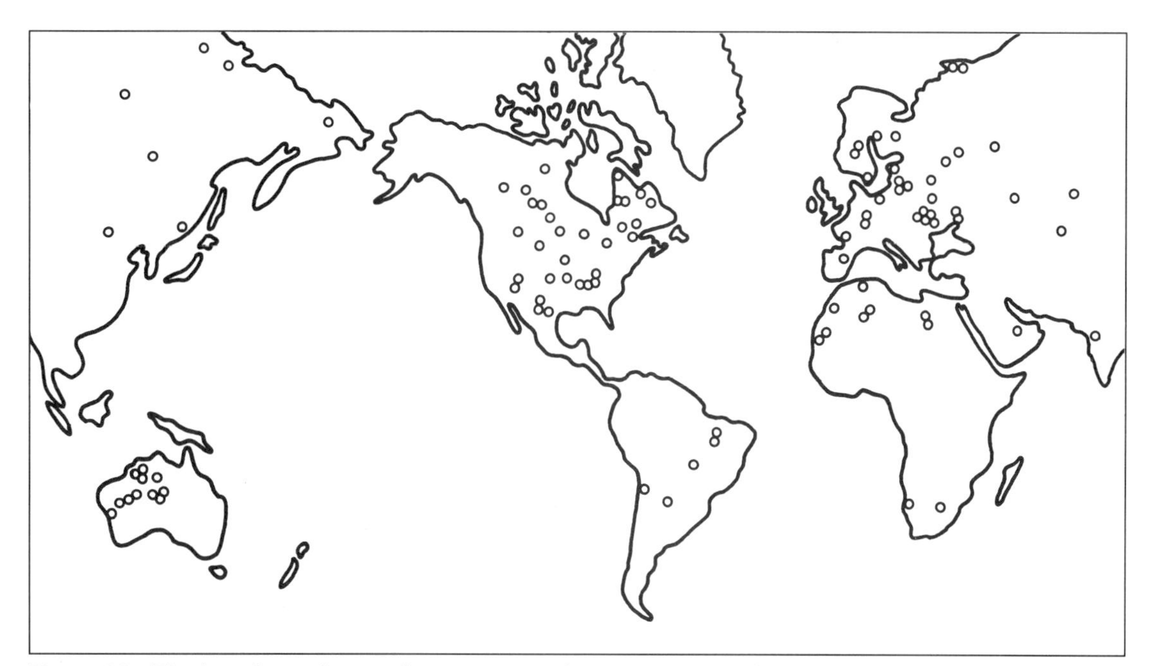

Figure 29 The locations of some known meteorite craters around the world.

In Quebec, Canada, the Manicouagan River and its tributaries form a reservoir around a roughly circular structure some 60 miles across (Figure 30). The structure is composed of Precambrian rocks that were reworked by shock metamorphism generated by the impact of a large celestial body. The impact crater was formed about 210 million years ago and coincides with the mass extinctions at the end of the Triassic period that eliminated most reptilian species and paved the way for the dinosaurs.

Figure 30 The Manicouagan impact structure, Quebec, Canada, from *Skylab* in 1973. Courtesy of NASA

The vast majority of ancient meteorite impacts have long been erased by the Earth's active erosional processes, including the action of wind, rain, glaciers, freezing and thawing, and plant and animal activity. The moon and

Mars retain most of their craters because they lack a hydrologic cycle, whose erosional forces have wiped out most impact structures on Earth. The forces of erosion have leveled the tallest mountains and gouged out the deepest canyons; no wonder most craters cannot escape these powerful weathering agents. The exception are craters in the deserts, which do not receive significant rainfall, or in the Arctic tundra regions, which remain unchanged for ages.

Apparently, very large craters that are over 12 miles in diameter and more than 2.5 miles deep are practically impervious to erosion. They escape erosion because the Earth's crust literally floats on a dense, fluid mantle. The process of erosion, whereby material is gradually removed from the continents, is delicately balanced by the forces of buoyancy that keep the crust afloat. Therefore, erosion can only shave off the upper 2 to 3 miles of the continental crust before the mean height of the crust falls below sea level.

Very large craters are usually deep enough so that even if the entire continent were worn by erosion, faint remnants would still remain. Craters of extremely large meteorite impacts might temporarily reach depths of 20 miles or more and expose the hot mantle below. The uncovering of the mantle in this manner would result in a gigantic volcanic explosion, releasing tremendous amounts of ash into the atmosphere that would exceed by far all the atmospheric products generated by the meteorite itself.

What appears to be the world's largest impact crater covers most of western Czech Republic and is centered near the capital city Prague. It is about 200 miles in diameter and at least 100 million years old. Concentric circular elevations and depressions surround the city, which is what would be expected if the Prague Basin were indeed a meteorite crater. Moreover, green tektites created by the melt from an impact were found in an arc that follows the southern rim of the basin. The circular outline was discovered in a weather satellite image of Europe and North Africa, and its immense size probably kept it from being noticed earlier.

Several methods can be utilized for detecting large impact craters that are invisible from the air. Seismic surveys could be used to detect circular distortions in the crust lying beneath thick layers of sediment. The disturbed rock usually produces gravity anomalies that could be detected with gravimeters. The fact that many meteorite falls are of the iron-nickel variety suggests that they could be detected by using sensitive magnetic instruments called magnetometers. The surface geology could also indicate areas where rocks were disturbed by the force of the impact or outcrop to form a large circular structure. The 10-mile-wide Wells Creek structure in Tennessee is in an area of essentially flat-lying Paleozoic rocks that were uplifted to form two concentric synclines (downfolded strata) separated by an anticline (upfolded strata).

METEORITES

The earliest reports of meteorite falls were made by the Chinese during the seventh century B.C. Chinese meteorites are rare, and to date no large impact craters have been recognized in China. The oldest meteorite fall, of which material is still preserved in a museum, is a 120-pound stone that landed outside of Ensisheim in Alsace, France, on November 16, 1492. The largest meteorite found in the United States is the 16-ton Willamette meteorite, which crashed to Earth sometime during the past million years. It was discovered in 1902 near Portland, Oregon, and measured 10 feet long, 7 feet wide, and 4 feet high.

One of the largest meteorites actually seen to fall was an 880-pound stone that landed in a farmer's field near Paragould, Arkansas, on March 27, 1886. The largest know meteorite find, named Hoba West, was located on a farm near Grootfontein, South-West Africa, in 1920 and weighed about 60 tons. The heaviest observable stone meteorite landed in a cornfield in Norton County, Kansas, on March 18, 1948. It dug a pit in the ground 3 feet wide and 10 feet deep.

Meteorite falls are a common occurrence, and every day thousands of meteoroids rain on the Earth. Occasionally, meteor showers involve hundreds of thousands of tiny stones (Figure 31). Luckily, most meteors burn up on their journey through the atmosphere, and their ashes contribute to the load of atmospheric dust. Upwards of 1 million tons of meteoritic material is produced annually.

When a meteor explodes near the end of its path through the atmosphere, it produces a bright fireball called a bolide. One such bolide produced the great fireball that flashed across the United States on March 24, 1933. Some bolides are bright enough to be seen during the daytime. Occasionally, their explosions can be heard on the ground and might sound like the sonic boom from a supersonic aircraft. It is estimated that every day thousands of bolides occur around the world, but most go unnoticed.

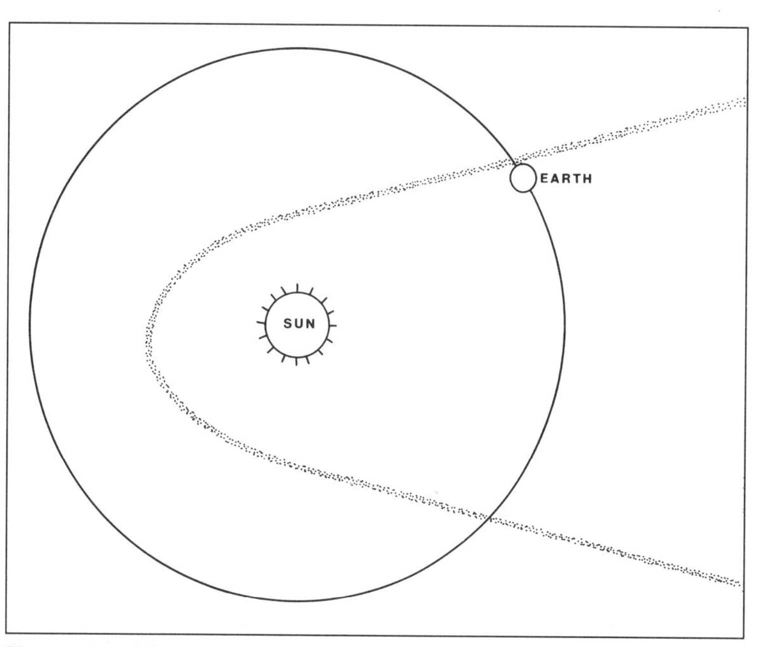

Figure 31 Meteor showers occur when the Earth intercepts the dusty path of a comet.

Over 500 major meteorite falls occur each year, most of which plunge into the ocean and accumulate on the seafloor. For the great majority of meteorites that land on the surface, the braking action of the atmosphere slows them down, so they only bury themselves a short distance. Not all meteorites are hot when they land because the lower atmosphere tends to cool them, and some might even have a layer of frost on their surfaces. Meteorites can also cause a great deal of havoc—many have crashed into houses.

The most easily recognizable meteorites are the iron variety, although they only represent about 5 percent of all meteorite falls. They are composed of iron and nickel along with sulfur, carbon, and traces of other elements. Their composition is thought to be similar to that of the Earth's iron core, and indeed they might have once made up the core of a large planetoid that disintegrated long ago. Due to their dense structure, iron meteorites have the best chance of surviving an impact, and most are found by farmers plowing their fields.

One of the best hunting grounds for meteorites happens to be on the glaciers of Antarctica, where the dark stones stand out in stark contrast to the white snow and ice. When meteorites fall on the continent, they are embedded in the moving ice sheets. At places where the glaciers move upward against mountain ranges, the ice evaporates, leaving meteorites exposed on the surface. Some of the meteorites that have landed on Antarctica are believed to have come from the moon and even as far away as Mars (Figure 32), when large impacts blasted out chunks of material and hurled them toward the Earth.

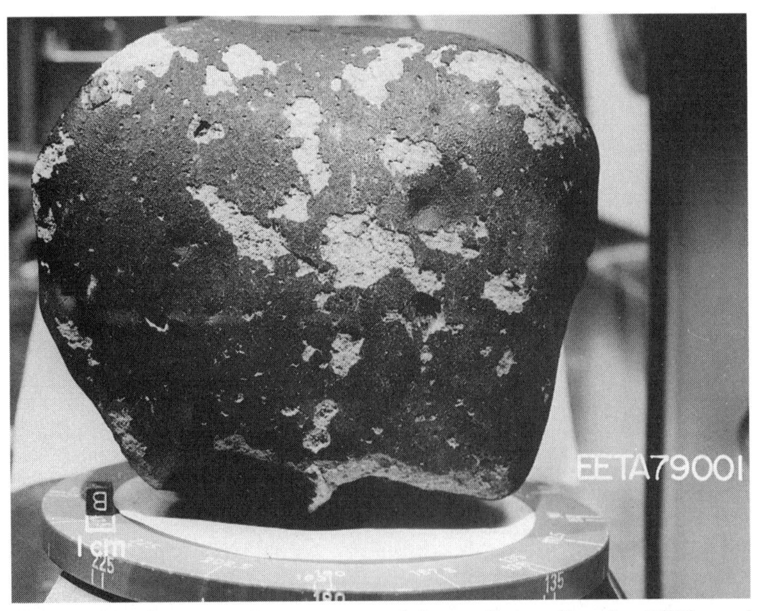

Figure 32 A meteorite recovered from Antarctica in 1981 and thought to be of possible Martian origin. Courtesy of NASA

Perhaps the world's largest source of meteorites is the Nullarbor Plain, an area of limestone that stretches for 400 miles along the south coast of Western and South Australia. The pale smooth desert plain provides a perfect backdrop for spotting meteorites, which are usually dark brown or black. Since very little erosion takes place, the meteorites are well preserved and are found just where they landed. Over 1,000 fragments from 150 meteorites that fell during the last 20,000 years have been recovered. One large iron meteorite, called the Mundrabilla meteorite, weighed more than 11 tons.

Stony meteorites are the most common type and make up more than 90 percent of all falls. But because they are similar to Earth materials and therefore erode easily, they are often difficult to find. The meteorites are composed of tiny spheres of silicate minerals in a fine-grained matrix. The spheres are known as chondrules, from the Greek *chondros* meaning "grain," and the meteorites themselves are called chondrites.

Figure 33 A North American tektite found in Texas in November 1985, showing surface erosional and corrosional features. Photo by E. C. T. Chao, courtesy of USGS

Most chondrites have a chemical composition believed to be similar to rocks in the Earth's mantle, which suggests that they were once part of a large planetoid that disintegrated eons ago. One of the most important and intriguing varieties of chondrites is the carbonaceous chondrites, which are among the most ancient bodies in the Solar System. They also contain carbon compounds that might have been the precursors of life on Earth.

STREWN FIELDS

Tektites, from the Greek *tektos* meaning "molten," are glassy bodies, ranging in color from bottle green to yellow brown to black (Figure 33). They are usually small, about the size of pebbles, although some have been known to be fist sized. Tektites come in a variety of shapes from irregular to spherical, including ellipsoidal, barrel, pear, dumbbell, or button-shaped. They also possess distinct surface markings that appear to have formed while the tektites solidified during their flight through the air.

Tektites have a distinct chemical composition unlike that of meteorites. It is much like the volcanic glass obsidian but with far less gas and water and no microcrystals, a characteristic unknown for any kind of volcanic glass. Tektites contain abundant silica similar to pure quartz sands like those used to make glass. Indeed, tektites appear to be natural glasses formed by the intense heat generated by a large meteorite impact. The impact flings molten material far and wide, and the liquid drops of rock solidify into various shapes while airborne.

Tektites cannot have originated outside the Earth because their composition closely matches that of terrestrial rocks. Their distribution over the Earth's surface suggests that they were launched at high velocity by a powerful mechanism, such as a major volcanic eruption or a large meteorite impact. However, terrestrial volcanoes are much too weak to produce the observed strewn fields of tektites that travel nearly halfway around the world. The largest of these is the Australasian strewn field, stretching across the Indian Ocean, southern China, southeastern Asia, Indonesia, the Philippines, and Australia (Figure 34). It comprises perhaps 100 million tons of tektites that are 750,000 years old. The Australasian tektites are roundish or chunky objects that show little evidence of internal strain.

Traces of ancient impact structures might also exist beneath the ocean or within the stratigraphic column. Mysterious glass fragments strewn over

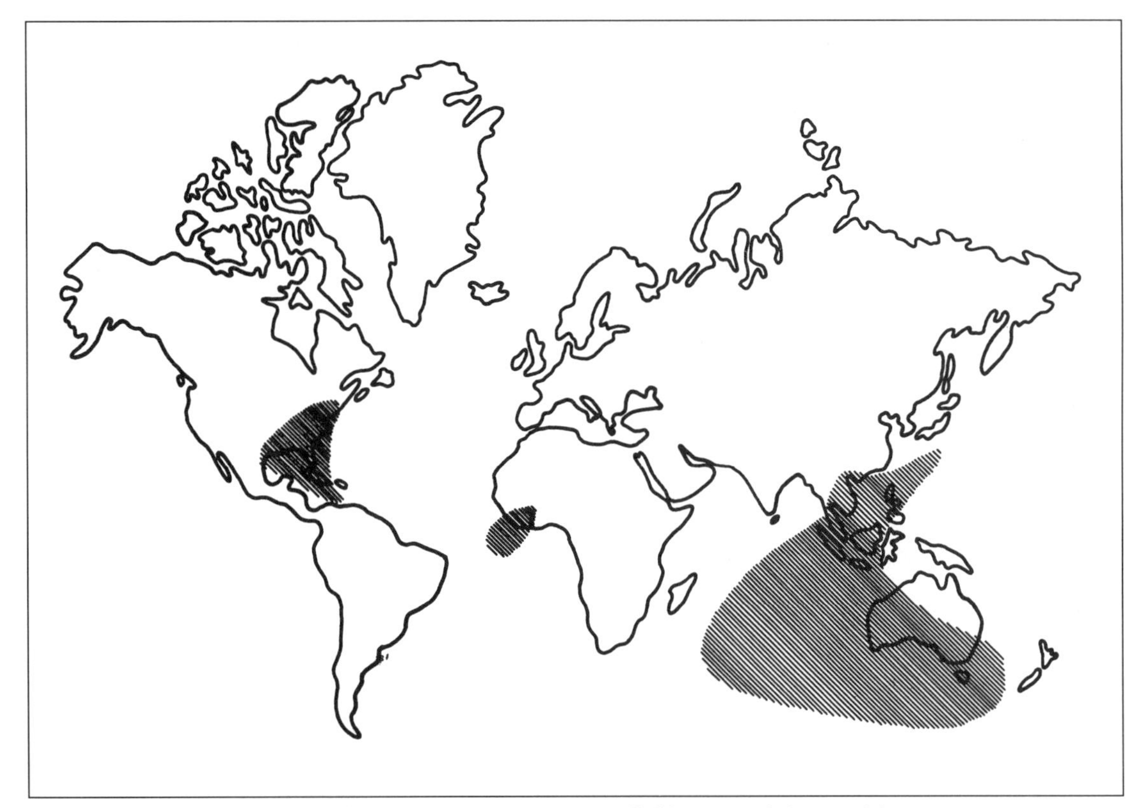

Figure 34 The distribution of tektites in major strewn fields around the world.

Egypt's Western Desert are believed to be melts from a huge impact that took place about 30 million years ago. Large, fist-sized, clear glass fragments are found scattered across the Libyan Desert. Analysis of trace elements indicates the glasses were produced by an impact into the desert sands. The 2-billion-year-old Vredefort structure in South Africa was identified as an impact structure on the basis of its impact melt and high iridium content.

Shocked metamorphic minerals believed to have originated from a massive impact 65 million years ago, when the dinosaurs became extinct, are strewn across western North America from Canada to Mexico. Shocked quartz and feldspar grains found in the Raton Basin of northeast New Mexico indicates that the impact occurred on land or the continental shelf because these minerals are rarely found in oceanic crust. The large size of the impact grains also suggests that the impact was nearby, perhaps within 1,000 miles of the strewn field.

In the cornfields of Manson, Iowa, lies a 22-mile-wide crater buried beneath 100 feet of glacial till. This site is ideal for the giant impact that created the North American strewn field. The rock type at the Manson site is the right composition and age and in the right location to account for the large size of the shocked quartz and feldspar grains in the western United States.

Because nearly three quarters of the Earth is covered by water, most meteorites land in the ocean, and several sites have been selected as possible marine impact craters. The most pronounced undersea impact crater is the 35-mile-wide Montagnais structure 125 miles off the southeast coast of Nova Scotia (see Figure 14). The crater is 50 million years old and closely resembles craters on dry land, but its rim is 375 feet beneath the sea and the crater bottom is 9,000 feet deep.

The crater was created by a large meteorite up to 2 miles wide. The impact raised a central peak similar to those seen inside craters on the moon. The structure also contained rocks that had been melted by a sudden shock. Such an impact would have sent a tremendous tidal wave crashing down on nearby shores. The crater is a likely candidate for the source of the North American tektites due to its size and location. However, its age indicates that the crater is several million years too young to have been responsible for the tektites. But the ocean is vast, and no doubt other craters will be found.

3

CAVES AND CAVERNS

Possibly no other geologic structure captures the popular imagination more than caves. They are, after all, our ancestral homes. Early geologists believed that the continents were undermined by huge subterranean caverns that collapsed and flooded the land with seawater, thereby producing the Great Flood. There is also a lot of misunderstanding about caves, and a great deal of superstition surrounds them.

Some caves hold extensive mazes where one could get lost forever. Many caves are homes to countless numbers of bats, which only exacerbates the misconceptions. Many people are simply afraid of caves or become claustrophobic due to the dark, enclosed spaces. On the other hand, spelunkers, or cavers, have an insatiable zeal for caves and are impatient to explore farther into the bowels of the Earth.

CAVE FORMATION

Caves are the most spectacular examples of the handiwork of groundwater. The dissolving power of water is well demonstrated by the formation of caves in soluble carbonate rock such as limestone. Although limestone caves are the most common, caves can also form in dolostone and gypsum.

Figure 35 Mammoth Hot Springs, Yellowstone National Park, Wyoming. Photo by W. H. Jackson, courtesy of USGS

Calcite is a calcium carbonate and the most common mineral in limestone. It is soluble in water that contains dissolved carbon dioxide. Another form of calcite, called travertine, is a common mineral found in caves and hot carbonate springs (Figure 35).

Rainwater filtering through sediment layers reacts with carbon dioxide to form a weak carbonic acid. In addition, if the overlying formation contains pyrite, the sulfur in the mineral could be oxidized by rainwater and converted into sulfuric acid. The acid is carried downward through cracks in the underlying rock layers and dissolves calcite or dolomite, the minerals that constitute limestone and dolostone. This action enlarges the fissures and eventually creates a path for more acidic water. Most of the dissolving occurs near the top of the water table, where a mixture of groundwater and water percolating from above dissolves rock more readily than either form of water by itself.

The wide variety in the shape and structure, or morphology, of cave passages is due to the chemical reactions that dissolve limestone and the position of the base level. The base level is the level at which water flows in an area and is influenced by the elevation of a nearby stream or river. Cave passages constructed near the top of the water table are affected by changes in base level. It also plays an important role in their orientation, even in areas where the landscape is more complex.

Different levels of passages in a cave develop at different times, when groundwater changes base levels. A lower level might begin to form below the water table, while an upper level forms near the top of the water table. Vertical shafts, up to several hundred feet deep, called domepits, connect the various levels. They develop relatively late in the cave-forming process, as rainwater trickles down toward the lower water table.

Cave passages can be affected by different geologic conditions, such as the existence of folds and faults at different parts of the cave. The greatest variety of cave morphology occurs where rock layers are folded. In some areas, rock layers were folded and pushed downward to form synclines. Other rock layers were folded and bulged upward to form anticlines. Many caves formed on the sides of synclines and anticlines have straight, parallel passages that run along the axis of the fold in the direction of the formation strike. Most caves form on only one level, but many located near the tops of mountains have several parallel levels of passages. These passages are connected by shorter ones that run along the slopes of the folded sedimentary beds.

The stratigraphy of horizontal rock layers is another important factor to cave development. The passage orientation and morphology is largely controlled by geologic, hydrologic and structural conditions. When a cave forms near the axis of a fold, a maze of passages might develop. The axes of synclines and anticlines are usually more fractured than the sides of the folds and transmit limestone-dissolving acidic water more easily. Joints are also important to the development of cave passages in areas where rock layers are not folded or tilted.

Caves also develop in the zone of seasonal water table fluctuation by the dissolution of limestone along joint planes. They form from underground channels that carry out water that seeps in from the water table. This creates an underground stream similar to how streams form on the surface by a breached water table. Caves also develop in sea cliffs (Figure 36) by the ceaseless pounding of the surf or by groundwater flow through an undersea limestone formation that is hollowed out as the water empties into the ocean.

Figure 36 Sea cave cut into siltstone. Chinitna district, Cook Inlet region, Alaska. Photo by A. Grantz, courtesy of USGS

Some caves were formed by tectonic movements of large rock masses that make up the Earth's crust. Other caves are made by weathering processes such as rock falls. Smaller caves are fashioned out of sandstone. Numerous caves are bored into the red sandstone formations around

Figure 37 A sinkhole in Minnekahta limestone, Weston County, Wyoming. Photo by N. H. Darton, courtesy of USGS

Moab, Utah. Houses built in natural and man-made caves are well insulated and maintain fairly uniform temperatures year round.

When groundwater percolates through limestone and other soluble materials, it dissolves minerals, forming cavities or caverns. When the land overlying these caverns suddenly collapses, it forms a deep sinkhole (Figure 37). At other times, the land surface might settle slowly and irregularly. The pits formed by dissolving soluble subterranean materials produce a pockmarked landscape called karst terrain. The major locations for karst terrain and caverns are mainly in the southeastern and midwestern United States, with Alabama and Florida containing the highest percentage. Half of Alabama is covered by limestone and other soluble rocks, producing thousands of sinkholes.

The karst terrain in the jungles of Mexico's Yucatan Peninsula gives birth to numerous underwater caves and sinkholes. The underlying limestone is honeycombed with huge caverns and tunnels, some several miles in length. Just like surface caves, the Yucatan caverns contain a profusion of stalactites and stalagmites. Creatures blinded by generations of unused eyesight live in the darkest recesses of these caves.

NATURAL BRIDGES

Many of the same processes that form caves also create natural bridges (Figure 38). They are narrow, continuous archways of rock that often span

Figure 38 A natural bridge in upper Carboniferous sandstone southwest of Douglas, Converse County, Wyoming. Photo by N. H. Darton, courtesy of USGS

a ravine or a valley. Rainbow Bridge near Lake Powell on the border between Utah and Arizona is the world's largest natural bridge.

Natural bridges are of complex origin and form by a variety of processes. They are the product of erosion and weathering of resistant rocks, such as sandstone or limestone, that contain layers that resist erosion in varying degrees. Some layers are hard and resist chemical and mechanical weathering, while others are easily weathered and eroded. If a resistant layer lies above a softer layer of rock, it forms a protective cap. When a vertical joint or fracture penetrates the softer rock and water flows through this structure, the softer rock erodes and undercuts the resistant capping layer.

Some natural bridges that formed in sandstone represent rock-shelter caves where part of the roof has collapsed, either by large blocks of rock breaking away or by a slower process of spalling, or chipping away grain by grain. When a narrow part of the roof is left intact, a bridge is formed. The roof collapse generally takes place in a section bounded by joints. Water from a surface stream might be captured in joints that penetrate the formation and flows through the rock. Eventually, the flowing water widens the joint and cuts below the uppermost layer of rock, resulting in a natural bridge.

Natural bridges in limestone or dolostone are formed by the chemical and mechanical weathering of an underlying, less-resistant layer of rock. Many of these bridges develop on narrow ridges. Generally, the opening beneath the bridge has been widened by solution weathering, with water percolating from the surface along a vertical joint. The passage beneath the bridge span is in effect a segment of a limestone cavern, formed by solution along a joint that has been exposed at either end as the ridge is narrowed by erosion.

Another type of natural bridge is formed when a large, detached block of rock falls or tilts so that it bridges the gap between two other blocks of rock. In limestone terrain, natural bridges are created in tunnels excavated by groundwater solutions, resulting in a collapse of the tunnel roof. Natural Bridge in Virginia (Figure 38A) is the most famous example of this type of bridge in the United States. Some natural bridges are fashioned by the lateral erosion of a stream flowing around and eventually through the rock. Other small bridges or arches are associated with sinkholes. Sea arches are created by wave action on limestone promontories with zones of differential hardness. Snow bridges often span crevasses in glaciers.

Differential weathering, which causes rock to erode at different rates due to differences in resistance, produces arches in arid regions like Utah, where they are especially plentiful (Figure 39). The arches were created partly by wind erosion of thick sandstone beds. Rainwater first loosens the sand near the surface, and wind removes the loose sand grains. Wind erosion then abrades the rock, cutting through it in a manner similar to sandblasting.

Figure 38A Natural Bridge, Rockbridge County, Virginia. Photo by J. K. Hillers, courtesy of USGS

LIMESTONE CAVES

Limestone is formed by biological and chemical precipitation of carbonaceous minerals dissolved in seawater. Carbonic acid, produced by the chemical

Figure 39 Gothic Arch in Navajo sandstone resting on Kayenta sandstone, Garfield County, Utah. Photo by H. E. Gregory, courtesy of USGS

reaction of water and carbon dioxide in the atmosphere, dissolves calcium and silica minerals from rocks on the surface to form bicarbonates. The bicarbonates enter rivers that empty into the ocean and are mixed with seawater and precipitate by biological activity and direct chemical processes. Organisms use the calcium bicarbonate to build their shells and skeletons, which are composed of calcium carbonate. When the organisms die, their skeletons fall to the bottom of the sea, where the calcium carbonate in the form of a calcite ooze builds thick deposits of carbonate rock (Figure 40).

The most common carbonate rock is limestone, which constitutes about 10 percent of all surface rocks. It is generally deposited by biological processes, as evidenced by marine fossils in limestone beds. Some limestone is chemically precipitated directly from seawater, and minor quantities precipitate in evaporite deposits. Dolostone resembles limestone, but is produced by the replacement of calcium with magnesium in limestone. The mineral is more resistant to acid erosion than limestone, which accounts for the impressive dolomite peaks of Europe.

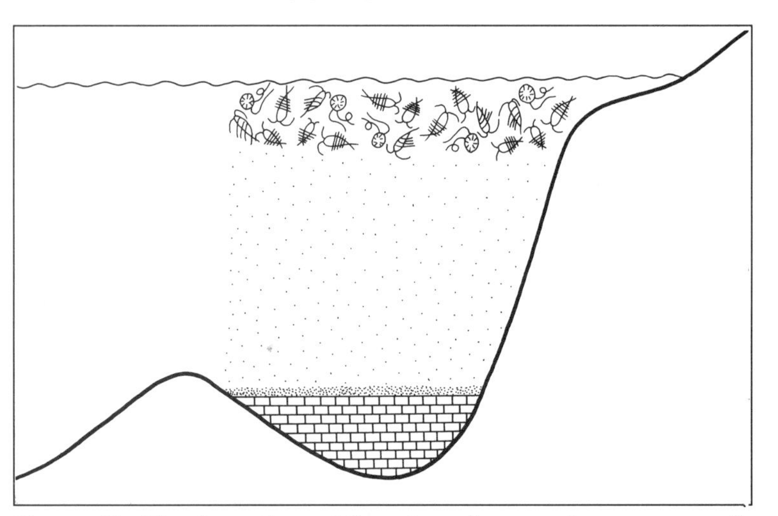

Figure 40 The formation of limestone from carbonaceous sediments deposited on the ocean floor.

Chalk is a soft, porous carbonate rock that erodes easily. Thick chalk beds were laid during the Cretaceous, which is how the period got its name, from the Latin word *creta* meaning "chalk." The thick chalk banks on the Suffolk coast of England have been wearing away by wave action for centuries. A major storm at sea can erode the tall cliffs landward as much as 40 feet. Sometimes the pounding of the surf punches a hole in the chalk to form a sea arch.

Over a lengthy period, groundwater dissolves large quantities of limestone, forming a system of tunnels, large rooms, and galleries. The difference in their shapes are due to the geology, hydrology, and structure of the rocks where the caves form. Anvil Cave in Georgia comprises nearly 13 miles of passages in an area of only 18 acres. Many caves possess more than one level, while some are entirely vertical. Ellison's Cave in Georgia has 10 miles of passages and over 1,000 feet of vertical drop, making it one of the deepest caves. The Mammoth Cave system in Kentucky is the longest in the world, with more than 300 miles of known passages spread among 6 levels. Most of the passages on each level are along bedding planes that separate different rock layers.

Figure 41 Hattin's Dome and the Temple of the Sun at Carlsbad Caverns National Park, Eddy County, New Mexico. Photo by W. T. Lee, courtesy of USGS

Carlsbad Caverns in southeast New Mexico (Figure 41) is formed from limestone that was once a large reef similar to Australia's Great Barrier Reef. The main portion of the cave is composed of massive limestone that remains much the same throughout the formation. The second part is layered and fashioned out of pieces of rock that eroded from the main part of the limestone and from overlying formations. The three largest caves formed at the area where the two formations meet, due to a layer of impermeable rock that lies between them. The many levels result from changes in groundwater base level during the cave's development.

LAVA CAVES

Lava is molten rock, or magma, that reaches the throat of a volcano or fissure and flows onto the surface with little or no explosive activity. Lava flows are generally tabular igneous bodies that are thin compared with their horizontal extent. Much of the shape of a lava flow is determined by the terrain upon which it flows. On flat plains, lava flows are usually horizontal, whereas on the slopes of volcanoes they might consolidate into considerable thickness. Lava flows greater than 300 feet thick are rare. In Hawaii, individual flows average 10 to 30 feet in thickness.

The magma from which lava is produced has a low viscosity, allowing volatiles and gases to escape with comparative ease. This gives rise to much quieter and milder eruptions such as those of the Hawaiian volcanoes. Lava is largely composed of basalt that is about 50 percent silica, dark and quite fluid. Pahoehoe, or ropy lavas, are highly fluid basalt flows produced when the surface of a flow congeals to form a thin plastic skin. As the melt beneath continues to flow, it molds and remolds the skin into billowing or ropy-looking surfaces. When the lavas eventually solidify, the skin retains the appearance of the flow pressures exerted on it from below.

Highly fluid lava moves rapidly, especially down the steep slopes of a volcano. The flow rate is also determined by the viscosity and the time it takes the lava to harden. Most lavas flow at speeds from a walking pace to about 10 miles per hour. Some lava flows have been clocked at only a snail's pace, while others run as fast as 50 miles per hour. Some very thick lavas creep along slowly for months or even years before solidifying.

If a stream of lava hardens on the surface and the underlying magma continues to flow away, a long tunnel, called a lava tube or cave, is formed. It can reach several yards across and extend for hundreds of yards. In exceptional cases the cave might extend up to 12 miles. Good examples of lava caves are found in the Modoc lavas in northeastern California and Craters of the Moon in Idaho (Figure 42). The caves might be partially or completely filled with pyroclastic material or sediments that washed in through small fissures. Sometimes the walls and roof of the lava caves are

Figure 42 The Bridge of the Moon lava tube, Craters of the Moon National Monument, Idaho. Photo by H. T. Sterns, courtesy of USGS

adorned with stalactites, and the floor is covered with stalagmites composed of deposits of lava.

In volcanic terrain in parts of Alaska, California, Oregon, Washington and Hawaii, volcanic-related subsidence is usually caused by local collapses above shallow tunnels. If the roof of a lava tunnel collapses, it leaves a circular or elliptical depression on the surface of the lava flow. A good example is a lava flow in New Mexico, where a collapse depression is nearly a mile long and 300 feet wide. Rilles

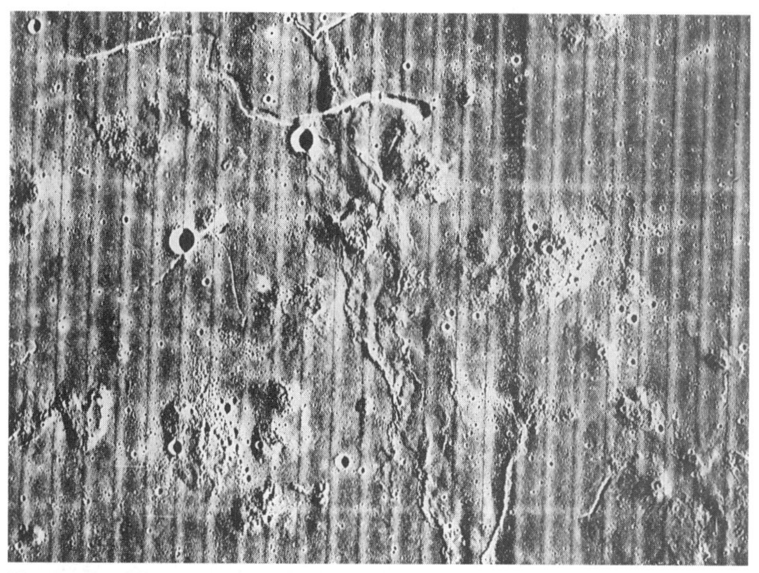

Figure 43 The heavily cratered Marius Hills region on the moon, showing domes, ridges and rilles on the lunar surface. Photo by D. H. Scott, courtesy of USGS

Figure 44 Meltwater from glaciers in Glacier National Park, Montana. Courtesy of National Park Service

on the moon are trenchlike valleys that resemble collapsed lava tubes (Figure 43). Lava tubes on Earth are very similar to rilles on the lunar surface as well as on the other planets and moons.

ICE CAVES

Iceland is a surface expression of the Mid-Atlantic Ridge and straddles both sides of the rift, where volcanic eruptions are quite frequent. Icelandic volcanism produces glacier-covered volcanic peaks up to 1 mile high. In 1918, an eruption under a glacier unleashed a flood of meltwater, called a glacial burst. In a matter of days, it released up to 20 times more water than the flow of the Amazon, the world's largest river. At least 13 similar underglacier eruptions have occurred during the last half of this century. Water gushing from such a glacier carves out an enormous ice cave. Geothermal heat beneath the ice creates a large reservoir of meltwater as much as 1,000 feet deep. A ridge of rock acts like a dam to hold back the water. When the dam suddenly breaks, the flow of water forms a channel under the ice up to 30 miles long or longer.

Meltwater flowing out of a glacier (Figure 44) often carves through the ice to form an ice cave that can be followed upstream for long distances.

Often the stream can be heard running beneath the glacier in crevasses slicing through the ice. The swift- flowing stream carries a heavy load of sediment that is deposited at the mouth of the glacier, forming glacial varves, alternating layers of silt and sand laid annually in a lake below the outlet of a glacier. Each summer when the glacial ice melts, turbid meltwater discharges into the lake and sediments settle out differentially, forming a banded deposit.

Long, sinuous sand deposits, called eskers, were formed out of glacial debris from outwash streams. They are winding, steep-walled ridges up to 500 miles in length but seldom exceed more than 2,000 feet wide and 150 feet high. Eskers are thought to have been created by streams running through tunnels beneath an ice sheet during the last ice age. When the ice melted, the old stream deposits remained standing as a ridge (Figure 45). Well-known esker deposits are found in Maine, Canada, Sweden and Ireland.

Drumlins are formed under the margin of a glacier and commonly occur in large numbers that face the advancing ice. They are composed of deposits of glacial till that produce elongated hillocks aligned in the same direction. Drumlins are tall and narrow at the upstream end of the glacier and slope to a low, broad tail. Drumlin fields might contain as many as 10,000 knolls. The hills appear in concentrated fields in North America, Scandinavia, Britain and other areas that were once covered with thick ice sheets.

Figure 45 An esker in Dodge County, Wisconsin. Photo by W. C. Alden, courtesy of USGS

Figure 46 Mirror Lake in Grand Caverns, Augusta County, Virginia. Photo by W. T. Lee, courtesy of USGS

Since the Little Ice Age, between 1430 and 1850, when global temperatures were about 3 degrees Fahrenheit cooler than they are today, outflow glaciers have been retreating, and their farthest advance is marked by terminal moraines composed of glacial deposits. Also, various limits of retreat can be identified by studying the growth of primitive plants called lichens on the rocks (lichenometry).

CAVE DEPOSITS

A cave's elaborate architecture is determined mainly by water filtering in from above. Rain and melting snow on the surface seep into the cave through thick deposits of soil and rock. As the water percolates downward, it picks up carbon dioxide from decaying plants to form a weak solution of carbonic acid. The acidic water trickles downward, dissolving limestone on its way. Upon reaching the surface of a cave, the water is exposed to the air and releases carbon dioxide, which precipitates calcite.

Limestone caves contain long "icicles" of calcite that grow by the precipitation of acidic groundwater seeping through the rock (Figure 46). A drop of water hanging from the ceiling of a cave when exposed to the air loses some of its acidity and can no longer hold calcite in solution. When the drop falls to the floor, a tiny calcite crystal is left behind. Over time, more drops gather at the same spot, and bit by bit the calcite grows downward from the ceiling, forming a stalactite; it is sometimes called dripstone because it is formed by dripping water containing dissolved calcium carbonate.

Water drops falling from the stalactite onto the cave floor still contain a small amount of dissolved calcite. When the drop hits the floor of the cave, it splatters and loses more acidity, precipitating another tiny calcite crystal. As more drops deposit calcite, a stalagmite grows upward toward the overhanging stalactite. Sometimes the two formations meet to form columns. The process is extremely slow, and it takes hundreds of years for stalactites and stalagmites to grow a single inch.

Some caves also contain exquisite twisting fingers of calcite or aragonite called helictites. They form like stalactites, but water drips through them too slowly for drops to form, creating contorted, branching cave deposits. The moisture reaches the tip of the helictite and evaporates, so crystals do not grow straight down but in curls and spirals. Aragonite, which is similar to calcite but has a different crystal structure, forms rough, needlelike helictites.

Other cave deposits include cave coral, created when water reaches the cave through a network of channels too small for drops to form. Instead, moisture is squeezed onto the cave wall, where it evaporates, leaving bumpy deposits that grow into various shapes, resembling popcorn, grapes, potatoes, and cauliflower. A drapery is formed when water droplets flow down a sloping cave ceiling, leaving trails of calcite that build up layer by layer. Sheets of calcite, called flowstone, are deposited when wide streams of water flow down cave walls, often constructing massive terraces.

CAVE ART

One of the earliest human species, called *Homo erectus*, appeared in Africa about 1.5 million years ago. By 1 million years ago, these people were present in southern and eastern Asia, where they lived until about 200,000 years ago. *Homo erectus* developed quite an elaborate culture, characterized by inhabiting caves and hunting game. Some populations might have been the first to use fire, possibly for hunting, cooking, and keeping warm.

Peking man was a variety of *Homo erectus* that lived in a large cave about 30 miles southwest of Beijing (Peking), China. People occupied this cave continuously for nearly a quarter of a million years, beginning about half a million years ago. Fossilized animal bones indicated that Peking man was an effective hunter. Fruits and grain were also a large part of his diet, as indicated by fossilized seeds found in the cave. As early as 400,000 years ago, Peking man could control fire and keep it burning, although it is not clear that he could ignite it; he probably relied on natural blazes started by lightning strikes. Indications that he utilized fire to cook his food is evidenced by quantities of charred seeds in the cave.

The Neanderthals are generally thought to be cave dwellers because most of their bones have been found there, due to the fact that caves preserved bones better than open sites. When not occupying caves, the Neanderthals lived in open-air sites, as indicated by hearths and rings made of mammoth bones and masses of stone tools normally associated with these people. The Neanderthals might have even made rock carvings and cave paintings. They buried their dead and placed offerings such as ibex horns and flowers in the graves.

It also has been postulated that Neanderthals practiced cannibalism. In Italy's Guattari Cave, where Neanderthals lived between 50,000 and 100,000 years ago, an adult male skull was found within a ring of stones from an apparent ritual of cannibalism. Human bones found at a cave in former Yugoslavia have long been viewed as the remains of a cannibal feast that occurred more than 50,000 years ago. What appears to be human cannibalism has emerged at other sites, including 6,000-year-old remains in France. The scientific community remains divided on whether cannibalism was practiced routinely and systematically at such locations or whether it occurred only in rare cases of imminent starvation.

Figure 47 A typical drawing of an animal on cave walls by primitive people.

Modern humans, called Cro-Magnon, might have originated as early as 100,000 years ago in the sub-Saharan region of Africa, where some of the oldest finds of modern man were discovered. They were named for the Cro-Magnon cave in France, where the first discoveries were made in 1868. Their appearance was markedly different from the stocky Neanderthals, and they shared the great majority of physical attributes of humans today.

Figure 48 Little Long House in overhanging ledge of Cliff House sandstone. Photo by C. H. Dane, courtesy of USGS

Cave paintings were a common form of human expression, and one cave in the French Pyrenees has walls containing over 200 human hand prints, most with missing fingers, dating to about 26,000 years ago. The fingers might have been destroyed by disease or infection or hacked off in some sort of ritual. The late ice age people tended to make elaborate and beautiful drawings and carvings of animals (Figure 47), especially those they did not eat, such as horses. Cave paintings in Brazil suggest that cave art began in the Americas almost at the same time it appeared in Europe and Africa.

Below numerous sandstone cliffs, such as those in the American Southwest, are adobe dwellings of the Anasazi Indians (Figure 48), who mysteriously vanished around 800 years ago, possibly due to a prolonged period of drought. Other Indians carved or painted on cave walls a variety of figures, including animals, called petroglyphs. Originally, the carvings were thought to be simple pieces of cave art. However, many carvings bear a certain relation to the sun's path across the sky, and are believed to be calendars.

4

CANYONS AND VALLEYS

Some of the most impressive scenery the planet has to offer is carved out of solid rock by water in motion. Erosion gouges out the deepest ravines, cuts down the tallest mountains, and obliterates most other geologic structures. Perhaps the best place to witness the power of erosion is the Grand Canyon of the Colorado River. The ocean bottom rivals the land for its rugged topography, and many ravines on the seafloor can hold several Grand Canyons. Some of the most monumental landforms are carved out of the earth by massive glaciers. Rivers and streams also play a major role in sculpting the landscape, providing an abundance of unusual geologic structures.

EROSION

Erosion is a geologic process that carves the landscape into jagged mountains and ragged canyons. The rise of active mountain ranges is matched by erosion, resulting in practically no net growth. No matter how pervasive mountain ranges formed by the forces of uplift are, they eventually lose the battle with erosion and are worn down to the level of the prevailing plain, with only their deep roots remaining to mark their existence.

CANYONS AND VALLEYS

The cores of the world's mountain ranges contain some of the oldest rocks on Earth. What was once buried deep beneath the surface has been thrust upward to great heights. Huge blocks of ancient granite that formed the interiors of the continents were pushed up by tectonic forces operating deep within the Earth and exposed by erosion.

Erosion also gouges out deep ravines in the hardest rock and has obliterated most geologic structures, including ancient man-made structures. It has carved out some of the most impressive geologic features the planet has to offer, from deep canyons and broad valleys to individual monuments sculpted from stone. Massive sandstone cliffs in the western United States have been slowly eroding for millions of years. Perhaps nowhere else on Earth is this process better demonstrated than the Grand Canyon of northern Arizona, which was carved out by the swift-flowing Colorado River, now a mere trickle of its former self.

Figure 49 Pinnacles of colorful Wasatch mudstone, Bryce Canyon National Park, Utah. Photo by P. L. Bradley, courtesy of USGS

Figure 50 Monuments and broad flats between them, Monument Valley, Navajo County, Arizona.
Photo by I. J. Witkind, courtesy of USGS

Canyonlands National Park in east-central Utah possesses a forbidding labyrinth of tributary canyons carved thousands of feet deep in beautiful red Permian sandstone. Bryce Canyon National Park in southern Utah displays pillars of mudstone that were carved out of the colorful Oligocene Wasatch Formation (Figure 49). Similarly colored sediments are responsible for the Painted Desert of Arizona and the Badlands of South Dakota, where short, steep slopes were eroded by numerous small streams, forming a unique drainage network.

Monument Valley on the border between Utah and Arizona in the Four Corners region contains both isolated and groups of monuments that tower 1,000 feet or more above the desert floor (Figure 50). A resistant cap rock preserved the sediments below, while the surrounding landscape eroded. A similar situation, on a broader scale, is exhibited on flattop mountains, where a remnant of the original peneplain, literally meaning almost a plain, is protected by a more resistant layer of sandstone. Many mesas, such as the Grand Mesa of western Colorado, the largest in the world, owe their existence to an upper layer of resistant sandstone or basalt.

Erosion rates vary, depending on the amount of precipitation, the topography of the land, the type of rock and soil material, and the amount of vegetative cover. In the past, erosion rates might have been higher than they are today. Continents of the early Earth were desolated places practically devoid of soil because water washed away nearly all the loose sediments.

The greatest source of sediments today are the towering Himalaya Mountains between India and China. Wind and water are slowly eroding the mountains into tiny grains that eventually wash into the Bay of Bengal by the Ganges and Brahmaputra rivers, which carry as much as 40 percent of the total sediment discharged by all rivers. The sediments are dumped into the adjoining basin to a depth of 12 miles.

In the Earth's early stages, the relief of the land was not nearly as great as it presently is. It took millions of years of mountain building and erosion to present us with a landscape of tall mountains and deep canyons.

TERRESTRIAL CANYONS

Some of the most magnificent examples of the power of erosion are canyons. They are generally found in arid or semiarid regions, where the effect of stream erosion is greater than that of weathering. Arizona's Grand Canyon (Figure 51) is a great gash in the Earth's crust 277 miles long, 10 miles wide on average, and over 1 mile deep. It is located at the southwest end of the Colorado Plateau, a relatively mountain-free expanse that stretches from Arizona north into Utah and east into Colorado and New Mexico.

Figure 51 The Grand Canyon from Torroweep Point, Coconino County, Arizona. Courtesy of National Park Service

Initially, the area surrounding the canyon was almost totally flat. Over the last 2 billion years, heat and pressure buckled the land into mountains that were later flattened by erosion. Again, mountains formed and were eroded and flooded by shallow seas. Afterward, the land was uplifted during a time when the Rocky Mountains were growing, between 80 and 40 million years ago.

The Grand Canyon is a relatively young feature in the Earth's crust. It was formed by forces of uplift and erosion as the Colorado River sliced its way through half a billion years of accumulated sediments and Precambrian basement rock. About 10 to 20 million years ago, the Colorado River began eroding layers of sediment, and its present course is less than 6 million years old. Much of the canyon was not carved out by piecemeal erosion grain by grain but by catastrophic landsliding that tore away entire canyon walls.

The canyon provides some of the best rock exposures on the North American continent. It slices through rocks that are hundreds of millions of years old and over 1 mile thick. Well-exposed sedimentary layers tell a nearly complete story about the geologic history of the area. On the canyon

Figure 52 The Precambrian Vishnu Schist on the bottom of the Grand Canyon. Photo by R. M. Turner, courtesy of USGS

wall is a unique angular unconformity separating the upper horizontal Plateau series from the older, tilted Grand Canyon series. On the bottom of the canyon lies the ancient original basement rock (Figure 52), upon which sediments were slowly deposited layer by layer.

Some of the best exposures of Precambrian rocks in the United States are the 1.8-billion-year-old metamorphic rocks on the bottom of the Grand canyon. Over a mile of sedimentary rocks overlie the bedrock of the canyon. The oldest of these rocks dates back to about 800 million years, leaving 1 billion years of history missing from the geologic record. During this time, the floor of the Grand Canyon was worn by erosion, creating a gap in time known as a hiatus.

Thick deposits of marine sediments were slowly laid on the Canyon floor. The continuous buildup of sediments caused the ancient seafloor to subside due to the increased weight. In a fraction of the time it took to deposit the sediments, a gradual upheaval brought them to their present elevation, while the Colorado River gouged out layer upon layer of rock, exposing the raw earth below. The Imperial Valley of southern California owes its rich soil to the Colorado River as it carved out the mile-deep Grand Canyon and deposited its sediments in the area to a depth of 3 miles.

SUBMARINE CANYONS

On the floor of the ocean lies a rugged landscape that rivals anything on land. Chasms plunge to depths that dwarf even the largest terrestrial canyons. Several oceanic canyons slice through the continental shelf beneath the Bering Sea between Alaska and Siberia. About 75 million years ago, continental movements created a broad shelf that rose 8,500 feet above the deep ocean floor. Many times the shelf was exposed as dry land. During the height of the last ice age about 18,000 years ago, the level of the ocean dropped some 400 feet. Terrestrial canyons cut deep into the shelf. When the ocean refilled at the end of the ice age, massive landslides and mudflows swept down steep slopes on the shelf's edge, gouging out 1,400 cubic miles of sediment and rock.

On the continental shelf off the eastern United States, a step on the ocean floor has been traced for nearly 200 miles. It appears to represent the former ice age coastline, now completely submerged. The massive continental glaciers that sprawled over much of the Northern Hemisphere held enough water to lower the sea by several hundred feet. When the glaciers melted, the sea level returned to near its present position. Submarine canyons carved into bedrock 200 feet below sea level can be traced to rivers on the continent. They were carved by rivers emptying into the sea, when the sea level was lowered dramatically during the last ice age (Figure 53).

Figure 53 The extended shoreline during the height of the last ice age.

The continental margin and ocean floor off eastern North America are cut by numerous submarine canyons (Figure 54). Such canyons on the continental shelves and slopes are quite similar to river canyons, and some rival even the largest terrestrial canyons. They are characterized by high, steep walls and an irregular floor that slopes continually outward. The canyons are up to 30 miles or more in length and have an average wall height of 3,000 feet. One of the largest submarine canyons, the Great Bahamas Canyon, has a wall height of 14,000 feet, over twice that of the Grand Canyon.

Some submarine canyons were carved out of the ocean floor by ordinary river erosion during a time when sea levels were much lower than they are today, and many have heads near the mouths of large rivers. Other submarine canyons extend to depths of over 2 miles, too deep to have had a terrestrial river origin. They were formed by undersea slides, which are a much more effective method of eroding the ocean floor. The slides move rapidly down steep continental slopes covered mainly with fine sediments swept off the continental shelves by these slides. The slides themselves consist of sediment-laden water that is heavier than the surrounding seawater. The turbid water moves swiftly along the ocean floor and effectively erodes the soft bottom material. These muddy waters, called turbidity currents, can move down the gentlest slopes, and transport huge portions of the ocean floor.

DEEP-SEA TRENCHES

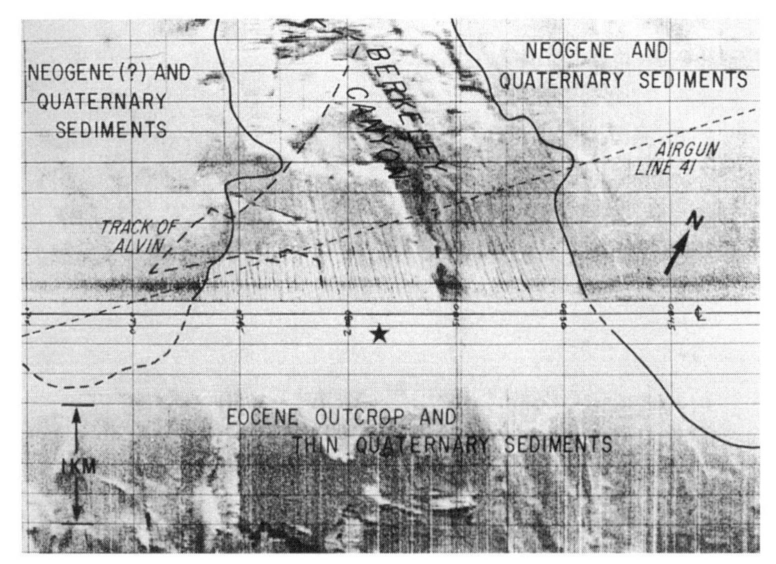

Figure 54 A sonograph of the lower continental slope off the Atlantic coast. Photo by N. P. Edgar, courtesy of USGS

The deepest spot in the world is the Pacific Mariana Trench (Table 5), which forms a long line northward from the island of Guam and reaches a depth of nearly 7 miles below sea level. The deep-sea trenches lie off continental margins and island arcs (Figure 55). They are regions of intense volcanic activity, producing some of the most explosive volcanoes on Earth. Volcanic island arcs fringe the trenches, and each has similar curves and similar volcanic origins. The trenches form in an arc because this is the geometric shape that occurs when a plane, such as a

TABLE 5 DIMENSIONS OF DEEP-OCEAN TRENCHES

Trench	Depth (miles)	Width (miles)	Length (miles)
Aleutian	4.8	31	2300
Japan	5.2	62	500
Java	4.7	50	2800
Kuril–Kamchatka	6.5	74	1400
Mariana	6.8	43	1600
Middle America	4.2	25	1700
Peru–Chile	5.0	62	3700
Philippine	6.5	37	870
Puerto Rico	5.2	74	960
South Sandwich	5.2	56	870
Tonga	6.7	34	870

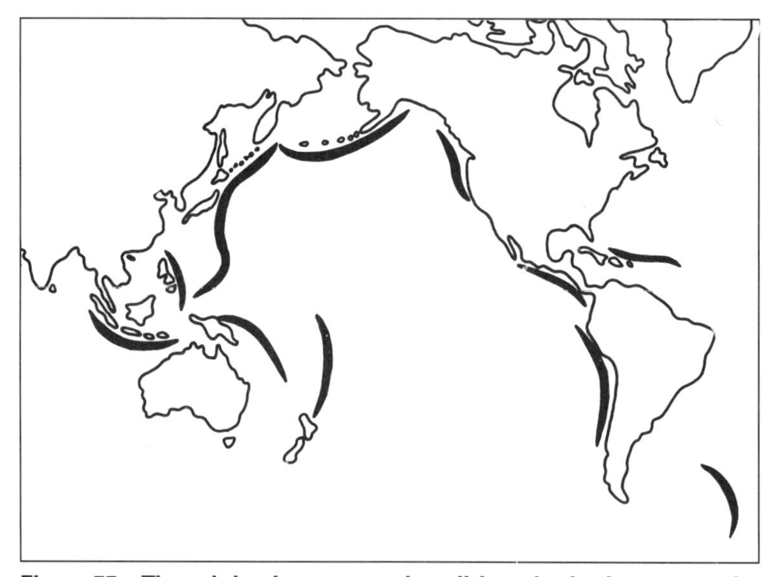

Figure 55 The subduction zones, where lithospheric plates enter the mantle, are marked by the deepest trenches in the world.

rigid lithospheric plate, cuts into, or subducts, a sphere, such as the mantle. The trenches are also sites of almost continuous earthquake activity deep in the bowels of the Earth. The earthquakes act like beacons that outline the boundaries of the plate.

A band of earthquakes might mark the earliest stages in the birth of a subduction zone that will eventually form a trench to the north and west of New Guinea in the western Pacific. Gravity there is lower than normal, a phenomenon that is expected over a trench due to the downward pulling of the ocean floor. Furthermore, a bulge in the crust to the south indicates that the edge of a slab of crust is beginning to be thrust downward into the Earth. If subduction is indeed being initiated, it could take another 5 to 10 million years before the process gets fully underway.

The early stages of subduction also might be occurring on the seafloor south of New Zealand. A massive earthquake took place in 1989 on the Marcquarie Ridge, an undersea chain of mountains and troughs running south from New Zealand and forming the boundary between the Pacific and Australian plates. As the Australian plate moves northwest in relation to the Pacific plate, ruptures occur along vertical faults that allow the plates to slip past each other. The plates also press together along dipping fault planes as they pass each other, creating compressional earthquakes. However, the dipping faults that flank the area have not yet connected to form one large fault plane, a necessary step before subduction commences. (For a further discussion of faults, see Chapter 7).

As the Pacific plate inches northwestward, its leading edge dives into the mantle, forming some of the deepest trenches on Earth. When a plate extends away from its place of origin at a midocean spreading center, it becomes thicker and colder, as more material from the asthenosphere below the plate adheres to its underside in a process known as underplating. Eventually, the plate becomes so dense it loses its buoyancy and can no longer remain on the surface. This causes it to sink into the mantle, and the line of subduction creates a deep-sea trench. As the subducted portion of the plate dives into the mantle, the rest of the plate, which might be carrying a continent on its back, is pulled along with it. This is the main driving mechanism for the drifting of the continents.

The deep-ocean trenches created by descending plates accumulate large amounts of sediments derived from the adjacent continents. The continental shelf and slope contain thick deposits of sediments washed off the continents. When the sediments and their content of seawater are caught between a subducting oceanic plate and an overriding continental plate, they are subjected to strong deformation, shearing, heating, and metamorphism. The sediments are carried deep into the mantle, where they melt to become the source of new molten magma for volcanoes that fringe the deep-sea trenches (Figure 56).

The magma rises to the surface in giant blobs called diapirs, from a Greek word meaning "to pierce." When the diapirs reach the underside of the lithosphere, they burn holes through it as they melt their way upward. After reaching the surface, the magma erupts on the ocean floor to create new

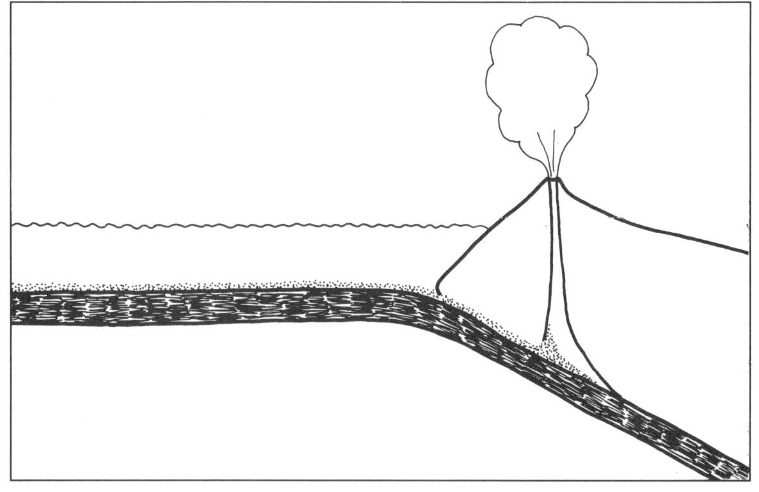

Figure 56 The subduction of the ocean floor provides new molten magma for volcanoes that fringe the deep-sea trenches.

Figure 57 The November 1968 eruption of Cerro Negro in west-central Nicaragua. Courtesy of USGS

volcanic islands. The volcanoes are very explosive because the magma contains a large quantity of volatiles and gases that escape violently when they reach the surface. The associated volcanic rock is called andesite, named for the Andes Mountains of Central and South America, which are well known for their violent eruptions (Figure 57).

The seaward boundaries of subduction zones are marked by deep trenches that are usually found at the edges of continents or along volcanic island arcs. Behind each island arc lies a marginal or back-arc basin, which is a depression in the ocean crust caused by plate subduction. Deep subduction zones like the Mariana Trench form back-arc basins, while shallow ones like the Chilean subduction zone off of South America do not. One classic back-arc basin is the Sea of Japan between China and the Japanese archipelago, which eventually will be plastered against Asia.

Figure 58 Yentna Glacier, Mount McKinley National Park, Alaska. Photo by N. Herkenham, courtesy of National Park Service

GLACIAL VALLEYS

The power of glacial erosion is well demonstrated by deep-sided valleys carved out of mountain slopes by thick sheets of flowing ice (Figure 58). Many valleys were buried by ice a mile or more thick during the ice ages. The glaciers descended from the mountains, spread across most of the northern lands, and like huge bulldozers destroyed everything in their way.

On mountains, large pits, called cirques, were gouged out by alpine glaciers flowing down the mountain peaks (Figure 59). Downstream from the foot of a glacier the surfaces of projecting rocks along the glacial valley floor look noticeably different from those high up on the sides of the valley. The higher rocks are rough and jagged, while those lower down are rounded, smooth, and covered with numerous parallel furrows, called glacial striae, pointing down the valley.

Figure 59 Athabaska Glacier, Alberta, Canada. Photo by F. O. Jones, courtesy of USGS

Figure 60 Polished and striated rocks near Cathedral Lake, Yosemite National Park, California. Photo by G. K. Gilbert, courtesy of USGS

The glaciers once extended far down the valleys, grinding rocks on the valley floors as the ice advanced and receded. In effect, a river of solid ice with rocks imbedded in it moved along the valley floors, grinding the rocks like a giant file as the glacier flowed back and forth over them. The advancing glaciers left marks on the valley floors as they ground their way down the mountainsides. Miles from existing glaciers are large areas of polished and deeply furrowed rocks (Figure 60), and heaped rocks called moraines marked the extent of former glaciers.

Some 200 surge glaciers that are heading toward the sea exist in North America. During most of its life, a surge glacier behaves normally, moving along at a snail's pace of perhaps a couple of inches a day. At regular intervals of 10 to 100 years, however, the glaciers gallop forward upwards of 100 times faster than their normal speed. One dramatic example is the Bruarjokull Glacier in Iceland, which advanced 5 miles in one year, at times achieving speeds of 16 feet an hour. For 85 years, Hubbard Glacier (Figure 61) has been flowing toward the Gulf of Alaska at a steady rate of about 200 feet per year. But in June 1986, the 80-mile-long river of ice surged ahead as much as 46 feet in one day.

Figure 61 Hubbard Glacier, Yakutat district, Alaska Photo by A. Post, courtesy of USGS

The routing of meltwater through a glacier can affect its movement. Glacial surges are therefore related to a glacier's internal drainage system. A surge often progresses along a glacier like a great wave, proceeding from one section to another like a gigantic caterpillar. Subglacial streams of meltwater help make the glaciers flow easily by acting like a lubricant. It is still a mystery why the glaciers surge, although they might be influenced by the climate, volcanic heat, and earthquakes. However, surge glaciers exist almost side by side with normal glaciers.

Antarctica has land features just like other continents except that major mountain ranges, deep canyons, high plateaus, and lowland plains are buried under a thick sheet of ice. Large flat areas beneath glaciers are thought to be subglacial lakes, kept from freezing by the interior heat of the Earth. The temperature a mile below the surface of the ice can be warmer than the temperature of the ice on top. Add to that the high pressures that occur at such depths, and liquid water can exist several degrees colder than its normal freezing point. The pools of water tend to lubricate the ice streams, helping them flow down the mountain valleys to the sea, where they calve to form icebergs. This allows ice streams up to several miles broad to glide smoothly along the valley floors.

Behind a wall of mountains that form the spine of Antarctica, called the Transantarctic Range, rivers of ice slowly flow outward and down to the sea on all sides. The ice escapes through mountain valleys to the ice-submerged archipelago of West Antarctica and to the great floating ice shelves of the Ross and Weddell seas. Dry valleys that run between McMurdo Sound and the Transantarctic Mountains (Figure 62) were originally gouged out by local ice sheets.

Figure 62 The upper part of Wright Dry Valley, Taylor Glacier region, Victoria Land, Antarctica. Photo by W. B. Hamilton, courtesy of USGS

The banks and interior portions of the ice streams are marked by deep crevasses. A glacial crevasse is a crack or fissure in a glacier, resulting from stress due to movement. They are generally up to 100 or more feet wide, 100 or more feet deep, and up to 1,000 or more feet long. The banks of glaciers are often flanked by deep crevasses, where they contact the walls of the glacial valley. Crevasses also run parallel to each other down the entire length of the ice streams, especially when the central portion of the glacier flows faster than the outer edges. Snow bridges occasionally span the crevasses, in some cases completely hiding them from view. Sometimes a stream of meltwater can be heard gurgling far below from open crevasses slicing through a glacier.

River Valleys

The United States is crisscrossed by over 3 million miles of rivers and streams. Most of the water on the continent is lost during floods. The rest is base flow, or the stable runoff of rivers and streams. The quickest route water takes back to the ocean is by river runoff. This is perhaps the most important part of the hydrologic cycle, which is the flow of water from the ocean, across the land, and back to the sea (Figure 63). As the surface water makes its way to the ocean, it transports sediments washed off the continents, which will ultimately be eroded to the level of the sea.

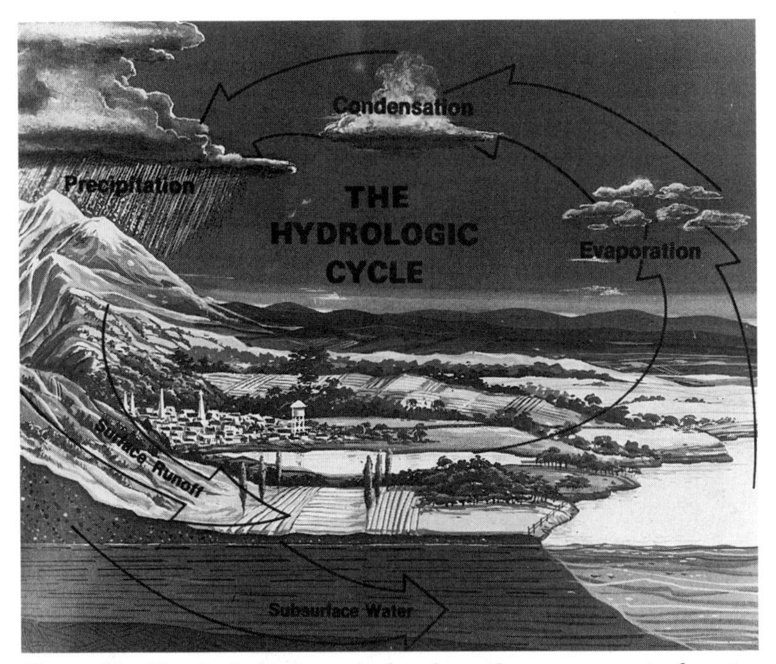

Figure 63 The hydrologic cycle involves the movement of water from the ocean, over the land, and back to the sea. Courtesy of USGS

On either side of a river channel is a floodplain, whose function is to carry excess water during floods. When rivers become clogged with sediment and fill their channels, they spill over onto the adjacent plain and carve out a new river course. In so doing, rivers meander downstream, forming thick sediment deposits in broad floodplains that can fill an entire valley. As a stream meanders across a floodplain, the greatest amount of erosion occurs on the outside of

Figure 64 Extensive flooding at Wilkes-Barre, Pennsylvania, from Hurricane Agnes in June 1972.
Photo by J. L. Patterson, courtesy of USGS

the bends, where the water moves faster, resulting in a steep cut bank in the channel. On the inside of the bends, the water slows down and deposits its suspended sediments.

Many of the nation's cities were originally located near watercourses, and the floodplains offered level ground for cultivation and construction. Unfortunately, failure to recognize the function of the floodplains has led to disastrous floods (Figure 64). Flash floods are the most intense form of flooding. They are local floods of great volume and short duration and are generally caused by severe thunderstorms over a relatively small drainage area. The discharges quickly reach a maximum and diminish almost as rapidly. The floodwaters frequently contain large quantities of sediment and debris collected as they sweep clean the stream channel. Flash floods can take place in almost any part of the country, but they are especially common in the mountainous areas and desert regions of the West.

Flash floods are particularly dangerous in areas where the terrain is steep, surface runoff rates are high, streams flow in narrow canyons, and severe thunderstorms are commonplace. Heavy rainfall in mountain regions causes sheets of water to run down steep mountainsides, picking up huge amounts of sediment that produces a swift, muddy flood. The floods can carry large boulders onto the floor of adjacent desert basins far beyond the base of the mountain range.

Riverine floods are caused by heavy precipitation over large drainage areas. They take place in river systems whose tributaries drain large geographical areas and encompass many independent river basins. Floods on large river systems might last from a few hours to many days. The flooding is affected by variations in the intensity, amount, and distribution of precipitation; the condition of the ground; the amount of soil moisture; and the quantity and type of vegetative cover.

The movement of floodwaters is controlled by river channel storage, changing channel capacity, and timing of flood waves. As the flood moves down the river system, temporary storage in the channel reduces the flood peak. As tributaries enter the main stream, the river increases in size farther downstream. Since tributaries are not the same size or spaced uniformly, their flood peaks reach the main stream at different times, smoothing out the peaks as the flood wave moves downstream.

Figure 65 Dendritic drainage pattern in an area underlain by Gila conglomerate, Gila County, Arizona. Photo by N. P. Peterson, courtesy of USGS

A drainage basin comprises the entire area from which a stream and its tributaries receive water. The Mississippi River and its tributaries drain a tremendous section of the central United States, reaching from the Rockies to the Appalachian Mountains. Furthermore, all tributaries emptying into the Mississippi have their own drainage area, forming a part of a larger basin. Every year, the Mississippi River dumps more than a quarter of a billion tons of sediment into the Gulf of Mexico, widening the Mississippi Delta and slowly building up Louisiana and nearby states. The Gulf Coast states from East Texas to the Florida panhandle were built up with sediments eroded from the interior of the continent and hauled along by the Mississippi and other rivers.

Streams and their valleys are joined into networks that display various types of drainage patterns, depending on the terrain (Figure 65). These patterns might be dendritic, resembling the branches of a tree, if the terrain is of uniform composition. Trellis drainage patterns are rectangular due to differences in the bedrock's resistance to erosion. Rectangular drainage patterns also occur if the bedrock is crisscrossed by fractures, forming zones of weakness that are particularly susceptible to erosion. Streams radiating outward in all directions from a topographic high form radial stream patterns.

An interesting drainage system was found buried beneath the sands of the eastern Sahara Desert, where there lies a vast network of river valleys as wide as the Nile Valley, along with smaller river channels, gravel terraces, desert basins and bedrock structures. It was one of the last great river systems on Earth, and its activity was much more intense than had been previously thought. Channels hundreds of thousands of years old twisted their way through valleys that are millions of years old but are now filled with thick deposits of sand. Scattered in the sand are Stone Age human artifacts, indicating that humans and life-sustaining water sources were once present in an area that is now totally uninhabitable.

5

MAJOR BASINS

Basins are the largest depressions in the Earth. If the oceans were completely drained of water, they would reveal basins with the deepest parts of seafloor lying several miles below the surrounding continental margins. The surface of the desiccated ocean would be crisscrossed by the longest mountain ranges and fringed in places by the deepest trenches. The entire planet might look something like the rugged surface of Venus, which apparently lost its oceans eons ago.

The Mediterranean Sea shows signs of having dried out on several occasions in the past, producing a huge empty pit in the Earth's crust. Terrestrial basins that filled with water became inland seas and glacial lakes. Regions where the Earth's crust is being stretched, such as the Great Basin of North America, are subsiding, and some areas like Death Valley have fallen far below sea level.

OCEAN BASINS

Most of the seawater that surrounds the continents is held in one great basin in the Southern Hemisphere, which branches northward into the Atlantic, Pacific, and Indian Oceans. The Arctic Ocean is a nearly enclosed sea that is connected to the Atlantic and Pacific by narrow straits. The oceans cover about 70 percent of the Earth's surface. Some 60 percent of the planet is covered by water over 1 mile deep, and the average depth is about 2.5 miles. Yet compared with the large size of the Earth this is merely a thin veneer. In the Pacific Basin, the ocean is up to 7 miles deep, and if Mount Everest, the world's tallest terrestrial mountain, were placed there the water would still extend over 1 mile above it.

The ocean is crisscrossed by vast undersea mountain ranges that are much more extensive than those on land. Active undersea volcanoes rise tens of thousands of feet off the ocean floor to create volcanic islands, some of which are actually the tallest mountains in the world. For example, Hawaii's Mauna Kea Volcano rises 32,000 feet from the ocean floor, exceeding the height of Mount Everest by about 3,000 feet. Some canyons on the ocean floor are several miles deep and rival the Grand Canyon in size and grandeur.

The vast majority of the volcanic activity that continually remakes the surface of the Earth takes place on the ocean floor (Figure 66), which is paved with black basalt, the most common volcanic rock. Almost all of this activity is confined to the margins of lithospheric plates. At convergent plate boundaries, where one plate is subducted under another, magma is formed when the lighter constituents of the subducted plate melt. The upwelling magma creates island arcs, mostly in the Pacific. These include Indonesia, the Philippines, Japan, the Kuril Islands and the Aleutians, the longest arc, extending over 3,000 miles from Alaska to Asia (See Figure 69).

Most volcanoes do not rise above the ocean surface and remain as isolated undersea volcanoes called seamounts. Since the crust under the Pacific Ocean is more volcanically active, it has a higher density of seamounts than the Atlantic or Indian Oceans. The number of undersea volcanoes increases with increasing crustal age and thickness. The tallest seamounts rise over 2.5 miles above the seafloor and are located in the western Pacific near the Philippine Trench, where the oceanic crust is more than 100 million years old. The average density of Pacific seamounts is between 5 and 10 volcanoes per 5,000 square miles of ocean

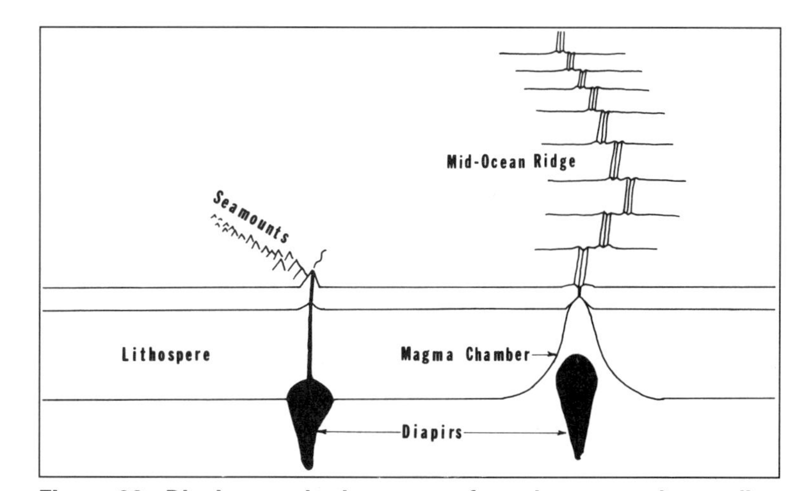

Figure 66 Diapirs supply the magma for volcanoes and spreading ridges on the ocean floor.

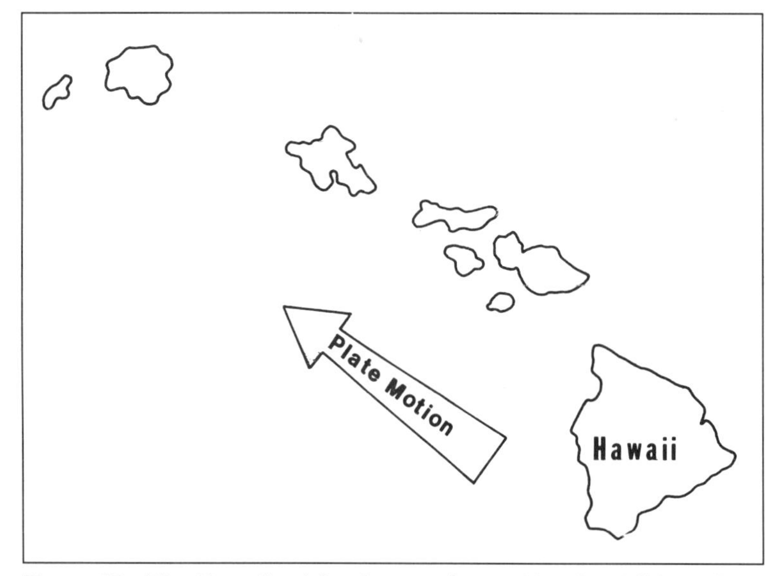

Figure 67 The Hawaiian Islands were formed by the drifting of the Pacific plate over a hot spot.

floor, a considerably greater occurrence of volcanoes than on the continents.

The passage of the Pacific plate over a mantle plume called a hot spot created the Hawaiian Islands; the youngest and most volcanically active island is Hawaii to the southeast, with progressively older islands whose volcanoes are extinct trailing off to the northwest (Figure 67). From there, coral atolls, like Midway Island, and shoals were formed by coral living on the flattened tops of volcanoes that were eroded below sea level, called guyotes (Figure 68). The atolls are rings of coral islands enclosing a central lagoon. They consist of reefs up to several miles across, and many have formed on ancient volcanic cones that have subsided beneath the sea, with the rate of coral growth exactly matching the rate of subsidence. Continuing from here are an associated chain of undersea volcanoes, called the Emperor Seamounts.

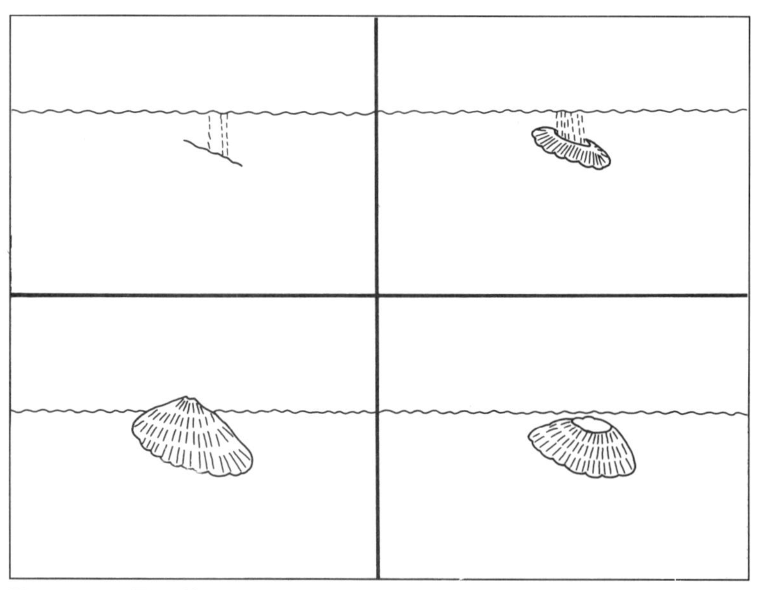

Figure 68 The life cycle of a guyot, a volcanic island that was eroded below sea level.

The Hawaiian Islands also lie parallel to the Austral Ridge and the Tuamotu Ridge that trend in the same general direction of many other seamounts. The islands and seamounts formed conveyor-belt fashion by the northwestward motion of the Pacific plate over a volcanic hot spot. The plate did not always travel in this direction, however. Over 40 million years ago, it followed a more northerly course. This course change, possibly as a result of a collision between India and Asia, which raised the great Himalayas, appears as a north-trending bend in the hot-spot tracks (Figure 69).

The Atlantic is a smaller, shallower ocean than the Pacific. It is split down the middle by the Mid-Atlantic Ridge, which manufactures new oceanic crust, as the continents surrounding the Atlantic Basin are spread farther apart. The Mid-Atlantic Ridge is the center of intense seismic and volcanic activity and the focus of high heat flow from the Earth's interior. Molten magma originating from the mantle rises up through the lithosphere and erupts on the ocean floor, adding new oceanic crust to both sides of the ridge crest.

As the Atlantic Basin widens, the continents surrounding the Atlantic Ocean separate at a rate of about an inch per year. The spreading of the ocean floor in the Atlantic causes the Pacific Basin to shrink in order to make more room. The Pacific is ringed by subduction zones (Figure 70) that are responsible for most of the geologic activity that surrounds the Pacific Ocean. Spreading ridges in the Pacific are also much more active than those in the Atlantic. Rapid spreading centers give rise to much lower relief on the ocean floor than slower ones because lava has less opportunity to pile up into tall heaps.

The oceanic plates thicken as they age, from a few miles after their formation at spreading centers to over 50 miles in the oldest ocean basins. The deepest abysses in the world are adjacent to continental margins, where the oceanic lithosphere is the oldest. The depth at which the oceanic crust sinks as it moves away from the midocean spreading ridges varies with its age. Crust that is 2 million

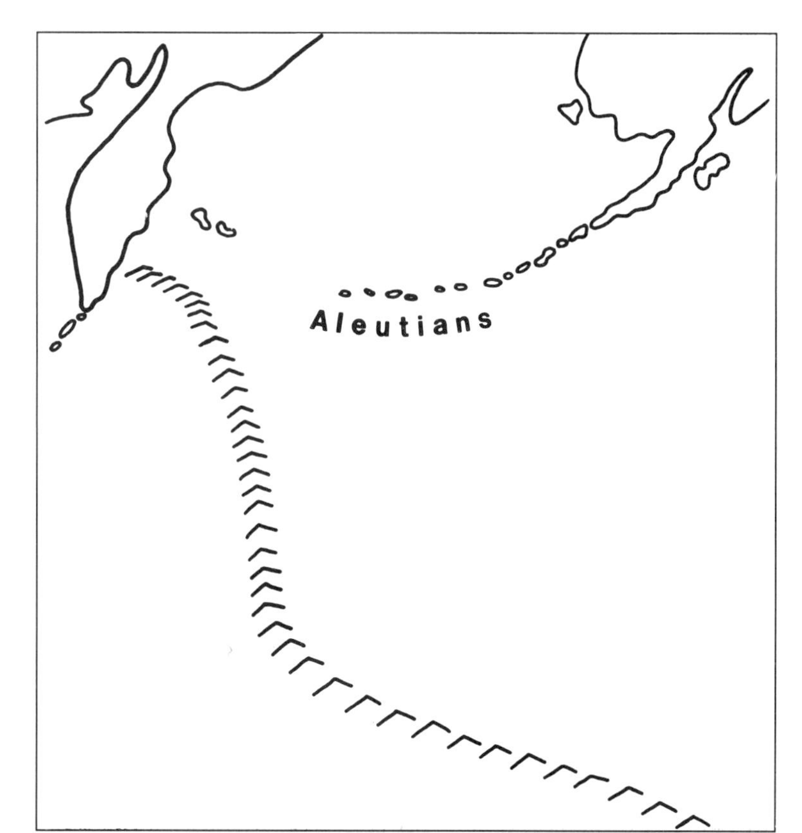

Figure 69 The Emperor Seamounts and the Hawaiian Islands in the North Pacific represent motions of the Pacific plate over a volcanic hot spot.

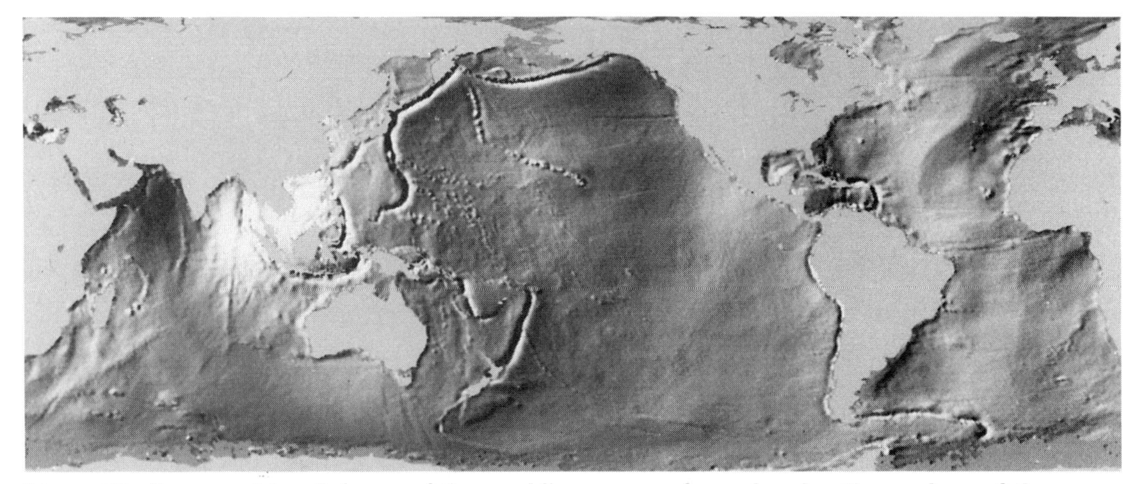

Figure 70 Topographic relief map of the world's ocean surface, showing the geology of the ocean floor. Courtesy of NASA

years old lies about 2 miles deep; crust that is 20 million years old lies about 2.5 miles deep; and crust that is 50 million years old lies about 3 miles deep.

The oceanic crust is composed of an upper layer of pillow basalts, formed when lava was cooled by seawater; a middle layer composed of a sheeted-dike complex, consisting of a tangled mass of feeders that brought lava toward the surface; and a lower layer composed of gabbros, which are coarse-grained basalts that crystallized slowly under high pressure. The ocean floor at the crest of the midocean ridge consists almost entirely of hard volcanic rock and acquires a thickening layer of sediments farther outward from the ridge crest.

If there were no bottom currents and only marine-borne sedimentation, an even blanket of material would settle onto the original volcanic floor of the world's oceans. Instead, the rivers of the world contribute a substantial amount of the material that ends up on the deep ocean floor. The largest rivers of North and South America empty into the Atlantic, which receives considerably more river-borne sediment than does the Pacific.

Since the Atlantic is smaller and shallower than the Pacific, its marine sediments are buried more rapidly and therefore are more likely to survive unscathed than are their Pacific counterparts. Moreover, the deep-ocean trenches around the Pacific trap much of the material that reaches its western edge, where it is subducted into the mantle. Thus, on the average, the Atlantic floor receives considerably more sediment than the Pacific floor, amounting to an inch per 2,500 years. In addition, strong near-bottom currents redistribute sediments in the Atlantic on a greater scale than they do in the Pacific.

Storms of intense currents occasionally sweep patches of ocean floor clean of sediments and deposit the debris elsewhere. On the western side of the ocean basins, periodic undersea storms skirt the foot of the continental rise and transport huge loads of sediment, dramatically modifying the seafloor. The energetic currents move about 1 mile per hour and can scour the ocean floor just as effectively as a gale with winds up to 45 miles per hour scours shallow areas near shore. The undersea storms seem to follow certain well-traveled paths, as indicated by long furrows in sediment on the ocean floor (Figure 71). The scouring of the seabed and deposition of thick layers of fine sediment result in much more complex marine geology than what would develop simply from a constant rain of sediments.

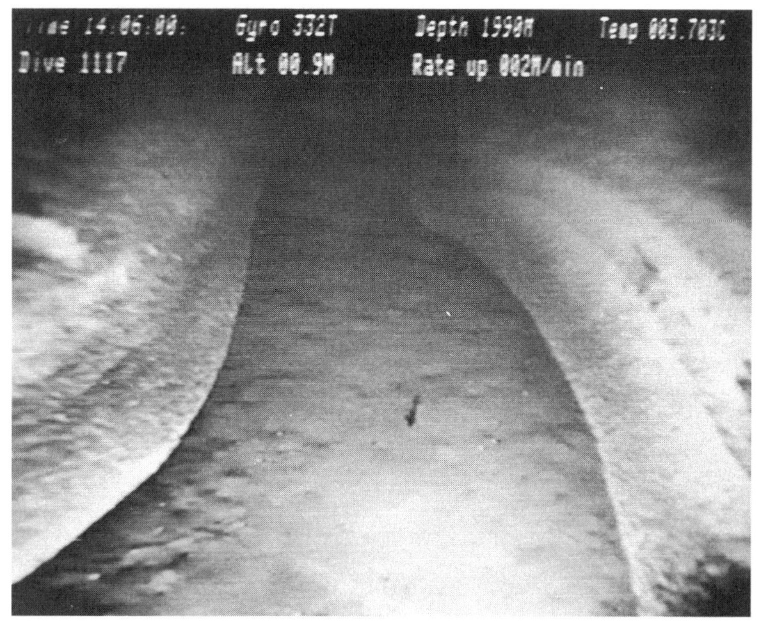

Figure 71 A wide flat furrow on the seabed of the Atlantic Ocean. Photo by N. P. Edgar, courtesy of USGS

INLAND SEAS

During the Cretaceous period, from about 135 million to 65 million years ago, the continents were flatter, mountain ranges were lower, and sea levels were higher. Seas invaded Asia, South America, Africa, Australia, and North

TABLE 6 HISTORY OF DEEP CIRCULATION IN THE OCEAN

Age (millions of years)	Event
50	The ocean could flow freely around the world at the equator. Rather uniform climate and warm ocean even near the poles. Deep water in the ocean is much warmer than it is today. Only alpine glaciers on Antarctica.

Age (millions of years)	Event
35–40	The equatorial seaway begins to close. There is a sharp cooling of the surface and of the deep water in the south. The Antarctic glaciers reach the sea with glacial debris in the sea. The seaway between Australia and Antarctica opens. Cooler bottom water flows north and flushes the ocean. The snow limit drops sharply.
25–35	A stable situation exists with possible partial circulation around Antarctica. The equatorial circulation is interrupted between the Mediterranean Sea and the Far East.
25	The Drake Passage between South America and Antarctica begins to open.
15	The Drake Passage is open; the circum-Antarctic current is formed. Major sea ice forms around Antarctica, which is glaciated, making it the first major glaciation of the Modern Ice Age. The Antarctic bottom water forms. The snow limit rises.
3–5	Artic glaciation begins.
2	An ice age overwhelms the Northern Hemisphere.

America. Thick layers of sediment were deposited into the basin of the interior seaway of North America (Figure 72) and are presently exposed as impressive sandstone cliffs in the western United States. Great deposits of limestone and chalk were laid down in the interior seas of Europe and Asia.

The interior of Australia was filled with a large sea during the Cretaceous, and sediments settling onto the floor of the basin were lithified into sandstone and shale, which were later heaved above sea level and exposed on dry land. Sitting in the middle of these sedimentary deposits are curious-looking boulders of exotic rock called dropstones, some measuring as much as 10 feet across. The boulders might have been rafted out to sea on slabs of drift ice. When the ice melted, the huge rocks dropped to the seafloor, where the impacts disturbed the underlying sedimentary layers.

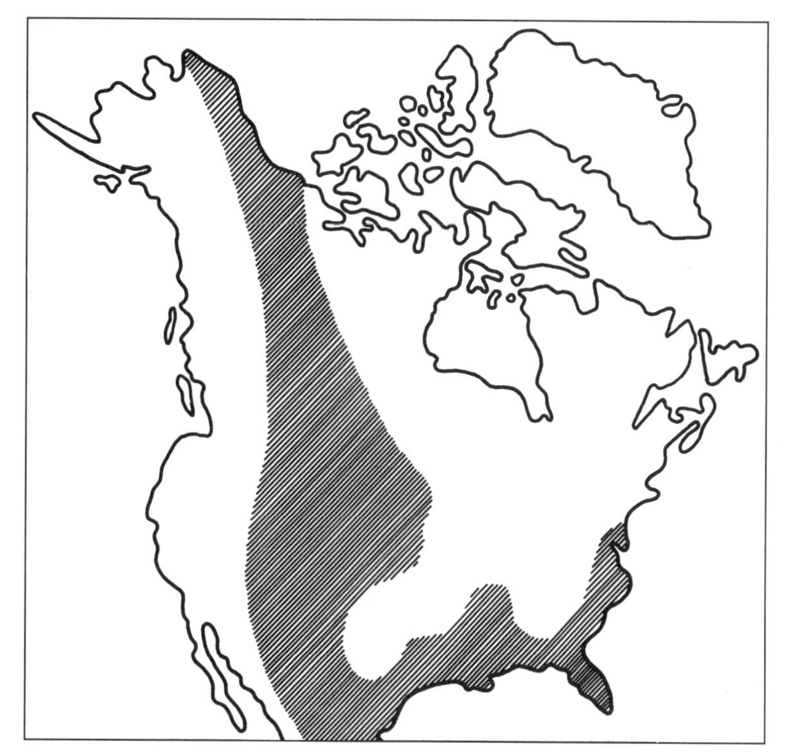

Figure 72 The Cretaceous interior sea in North America.

By the time the Cretaceous came to a close, three new bodies of water had formed due to continental drift: the Atlantic, Arctic and Indian oceans. The inland seas filled with sediments, and subsequent uplifting drove out the waters and left behind salt lakes. Great Salt Lake in Utah is only a remnant of a once vast inland sea. During a long wet period between 12,000 and 6,000 years ago, it expanded several times its current size and occupied the nearby salt flats (Figure 73). Africa's deserts were also dotted with several large lakes. Lake Chad, lying in central Africa on the border of the Sahara Desert, swelled to over 10 times its present size.

The lowest place on the Earth's surface is the Dead Sea, which is 1,300 feet below sea level. It is located in the Syrian Desert on the border between Israel and Jordan (Figure 74). The sea lies on the Syrio-African rift, formed by a transform fault, where two plates have been sliding past each other for the last 10 million years. The relative motion of the plates has stretched the crust, causing it to subside. For thousands of years, a stream carrying salts leached from surface rocks flowed south through the Jordan Rift Valley and emptied into the Dead Sea. Since there is no outlet, the inflowing water evaporates into the desert air, leaving salts to accumulate in the lake. As a result, the Dead Sea is not only the world's deepest salt lake, at about 1,000 feet deep, but also the world's saltiest. with an average salinity eight times greater than the ocean.

Since the 1960s, diversion of fresh water from rivers feeding the Dead Sea for irrigation of agricultural lands has substantially lowered the level of the lake, making its surface water saltier and much denser. This made

Figure 73 South end of the Great Salt Lake Desert, Utah, was part of a large lake at the end of the ice age. Photo by C. D. Walcott, courtesy of USGS

for a highly unstable condition, with a heavy, salty layer of water overlying a layer of lighter water on the bottom. This caused the lake to overturn, with the surface water exchanging places with the deeper water. The changes in salinity had a major impact on the Dead Seas's sparse biota, and after the overturn the lake was almost devoid of life.

Another inland sea, the Mediterranean, appears to have almost completely dried up 6 million years ago, and the seafloor became a desert basin over 1 mile below the surrounding continental plateaus. The adjacent Black Sea, which is 750 miles long and 7,000 feet deep, might have had a similar fate. Like the Mediterranean, it is a remnant of an ancient equatorial body of water called the Tethys Sea, which separated Africa from Eurasia and connected the Atlantic Ocean with the Indian Ocean (Figure 75). A collision of the African plate with Europe and Asia 20 million years ago squeezed out the Tethys, resulting in a long chain of mountains and two major inland seas. One was the ancestral Mediterranean and the other was a composite of the Black, Caspian, and Aral seas, called the Paratethys Sea, which covered much of eastern Europe.

About 15 million years ago, the Mediterranean and the Paratethys separated, and the Paratethys became a brackish sea, much like the Black Sea today. The disintegration of the great inland waterway was closely associated with the sudden drying up of the Mediterranean. In a brief moment in geologic time, the Black Sea became practically a dry basin. Then during the last ice age, it refilled again and became a freshwater lake. The brackish and largely stagnant sea that occupies the basin today has evolved since the beginning of civilization, about 10,000 years ago.

In the Salton Sea, the largest lake in California, the crust is thinning as blobs of magma rise toward the surface, resulting in a hot, fractured crust, whose salt-laden waters are tapped for their geothermal energy. The brine is eight times saltier than seawater and aids in dissolving metals in the sediments. The metals precip-

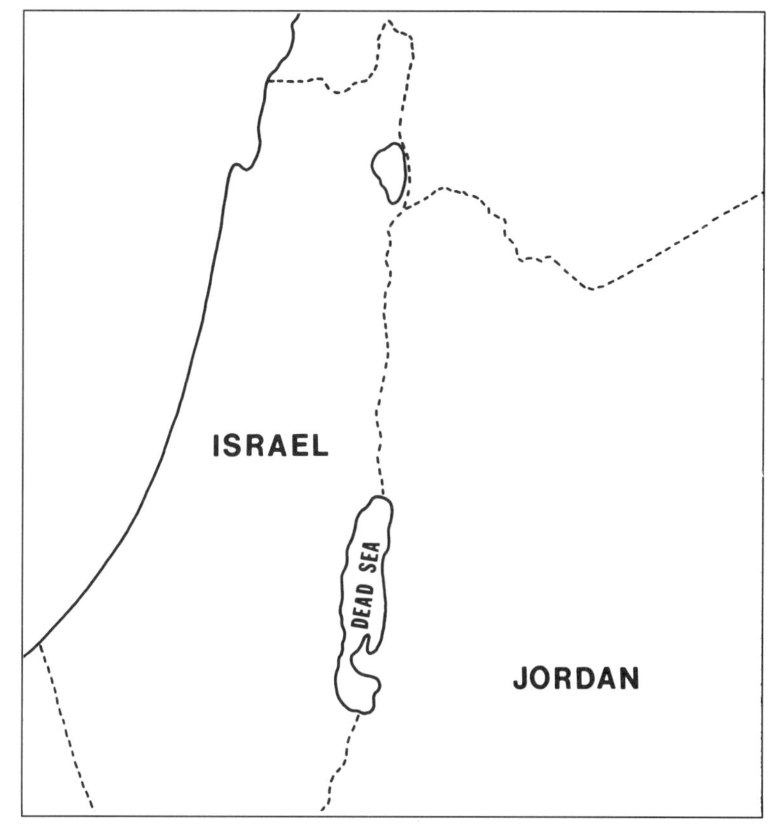

Figure 74 The location of the Dead Sea.

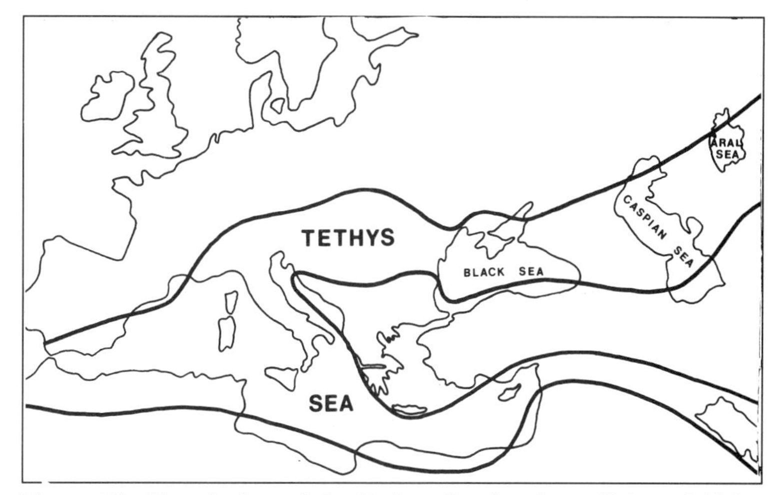

Figure 75 The closing of the Tethys Sea by the collision of Africa with Eurasia about 20 million years ago.

Figure 76 The Nile River delta, viewed from the Space Shuttle.
Courtesy of NASA

itate out of solution and are concentrated in fractures in the rock. The mineralization is similar to that found in rifts that were opening more than 600 million years ago.

DESICCATED BASINS

The Mediterranean Sea is nearly a completely enclosed basin, whose abyssal plains lie more than 10,000 feet deep. The basin holds almost 1 million cubic miles of seawater. The evaporation rate is extremely high, and nearly 5 feet of the water's surface evaporates every year, amounting to approximately 1,000 cubic miles of seawater. Only about one tenth of the loss is compensated by the influx of fresh water from rivers. The rest must be made up by seawater flowing in from the Atlantic Ocean through the narrow Strait of Gibraltar. The high salinity content makes the water heavier than normal seawater, causing it to sink to the bottom. Eventually, higher saline water will fill the entire basin.

The Mediterranean Basin was apparently cut off from the Atlantic Ocean 6 million years ago, when an isthmus was created at Gibraltar by the northward movement of the African plate, forming a dam across the strait. Nearly 1 million cubic miles of seawater evaporated, almost completely emptying the basin over a period of about 1,000 years. The dry basin was over 1 mile below the continental shelf and probably looked like a much larger version of Death Valley.

Rivers draining into the desiccated basin gouged out deep canyons. A deep sediment-filled gorge follows the course of the Rhone River in southern France for more than 100 miles and extends to a depth of 3,000 feet below the surface where the river drains into the Mediterranean Sea. Under the sediments of the Nile Delta (Figure 76) is buried a mile-deep canyon that can be traced as far south as Aswan 750 miles away and is comparable in size to the Grand Canyon of Arizona.

On the floor of the Mediterranean Sea is an array of salt domes several miles in diameter and up to thousands of feet high. They formed when salt diapirs buried below the seabed forced their way to the surface. Their presence indicates that extensive salt deposits underlie the floor of the Mediterranean. The salt deposits are interbedded with windblown sediments, which indicates the former presence of a dry basin. The evaporite deposits are about a mile thick and formed when the entire water column completely evaporated several times over a period of about 1 million years.

When Gibraltar subsided, a spectacular waterfall disgorged seawater at a rate of 10,000 cubic miles a year, requiring several centuries to refill the basin. The waterfall would have been 100 times larger than the mile-wide Victoria falls of southern Africa, one of the greatest falls on Earth, and 50 times higher than Niagara Falls of North America. The refilling of the Mediterranean Sea lowered the global sea level by as much as 35 feet. The mass of the water pressing down on the basin was comparable to the weight of the great ice sheet that spread across Europe during the last ice age.

Thick deposits of salt lining the seabed of the Gulf of Mexico might indicate that it also completely dried out around 140 million years ago. Salt domes similar to those in the Mediterranean Sea are common along the Gulf Coast, where many oil fields have been located around the domes. The North Sea and the Red Sea might also have experienced a similar fate and show signs of having evaporated because, like the Mediterranean Sea, they are floored with thick layers of salt.

GLACIAL LAKES

A large portion of the landscape in the northern latitudes was sculpted by massive ice sheets that swept down from the polar regions during the last ice age. The glaciation was so pervasive that glaciers 2 miles or more thick enveloped much of upper North America and Eurasia. In some places, the crust was scraped completely clean of sediments, exposing the raw basement rock below and erasing the entire geologic history of the region.

Around 13,000 years ago, a gigantic ice dam lying on the border between Idaho and Montana held back a huge lake hundreds of miles wide and up to 2,000 feet deep. When it suddenly burst, the waters gushed toward the Pacific Ocean, and along their way they carved out one of the strangest landscapes the planet has to offer, known as the Scablands (Figure 77). Lake Agassiz formed in a bedrock depression at the edge of the retreating ice sheet in southern Manitoba, Canada. It was a vast reservoir of meltwater that was much larger than any of the existing Great Lakes.

Figure 77 The Scablands of the northern end of the upper Grand Coulee River, eastern Washington.
Photo by F. O. Jones, courtesy of USGS

When the North American ice sheet retreated at the end of the last ice age, most of its meltwater flowed down the Mississippi River. Sometimes, huge lakes of meltwater trapped beneath the ice sheet broke free and rushed down river valleys in torrents to the Gulf of Mexico and the Atlantic Ocean equal to the flow of several Mississippi Rivers. While flowing under the ice, water surged in vast turbulent sheets scouring deep grooves in the surface, forming steep ridges carved out of bedrock. Several times, massive surges of meltwater broke loose to further gouge the landscape.

When the ice sheet retreated beyond the Great Lakes, which were themselves carved out by glaciers, the meltwater took a separate route down the St. Lawrence River. The cold waters entered the North Atlantic, initiating a return to ice age conditions and a pause in the melting known as the Younger Dryas, named after an Arctic wildflower. Also during this time, the high velocity water of Niagara River Falls (Figure 78) began cutting its gorge and has traversed over 5 miles northward since the ice sheet began to retreat, cutting into the bedrock at a rate of up to 3 feet per year.

Glacial lakes, dotting much of Canada and northern United States, were developed from major pits excavated by the glaciers. Smaller lakes were formed when large blocks of ice buried by glacial outwash sediments melted, forming depressions called kettles (Figure 79). The depressions are nearly circular, ranging up to 10 miles or more in diameter and up to 100 feet or more deep. Not all kettles contain water, however, and some, called dry kettles, might hold a stand of trees that gradually fall off below ground level toward the center.

The largest of the glacial lakes are the Great Lakes, bordering between the United States and Canada. The lakes presently receive huge quantities of sediments derived from the continent, and the constant buildup gradually makes them shallower. In the future the lakes will completely fill with sediment and become dry, flat, featureless plains, until the ice sheets return once again to scour out their basins.

Figure 78 Niagara Falls created as the Niagara River flows from Lake Erie northward into Lake Ontario. Courtesy of NASA

THE GREAT BASIN

The Basin and Range province is a 600-mile-wide region that includes the Great Basin area of Nevada and Utah. It covers southern Oregon, Nevada, western Utah, southeastern California, and southern Arizona and New Mexico (Figure 80). The Great Basin is a 300-mile-wide

Figure 79 A kettle hole in gravels near the terminus of Baird Glacier, Thomas Bay, Petersburg district, southeastern Alaska. Photo by A. F. Buddington, courtesy of USGS

closed depression formed by the stretching and thinning of the crust. The crust in the entire Basin and Range province is actively spreading apart because of forces originating in the Earth's mantle. These are the same forces that raised the Rocky Mountains during the Laramide orogeny, or episode of mountain building. The dominant element of the deformation is extension along a line running roughly northwest to southeast.

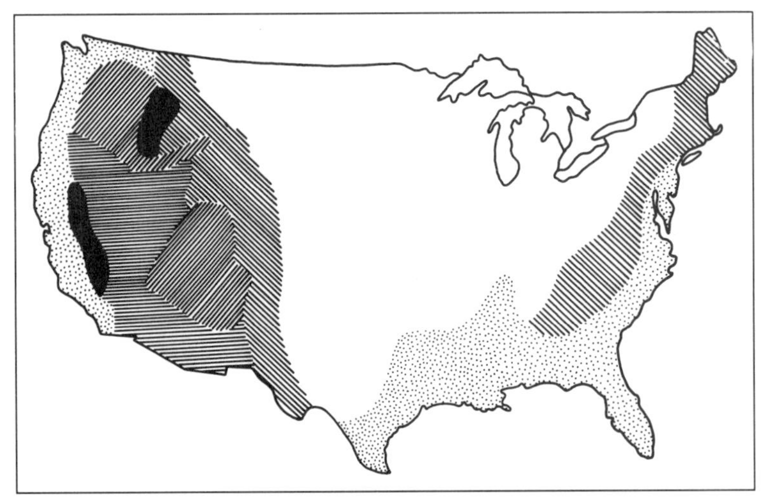

Figure 80 Physiography of the United States. Horizontal hatching represents the Basin and Range province.

As the crust continues to spread apart, some blocks sink to form grabens, which are fault-bounded valleys. Between the grabens are ridges called horsts, which are uplifted fault blocks. About 20 horst-and-graben structures extend from the 2-mile-high cliffs of the Sierra Nevada in California to the still-rising Wasatch front, a major fault system running roughly north-south through Salt Lake City, Utah.

The horst-and-graben structures trend nearly perpendicular to the movement of blocks of crust as they spread apart. This movement is part of the "missing motion" between the Pacific and the North American plates. The Pacific plate is sliding northwestward past the North American plate at about 2 inches per year. The relative motion of these two enormous sections of the Earth's crust is responsible for the deformation of the crust and the ongoing tectonic activity taking place across the western United States. This includes the expansion of the Basin and Range province, the horizontal slippage along the San Andreas Fault of California, and the uplifting of California's coastal ranges.

The San Andreas Fault absorbs between 60 and 80 percent of the relative motion between the Pacific and North American plates. The deformation of California's rugged coastal ranges takes up about 10 percent of the motion, causing intense folding and thrust faulting. The Garlock Fault in California (Figure 81) is a major east-trending fault, whose left-lateral movement combined with the right-lateral movement of the San Andreas Fault is causing the Mojave Desert to move eastward with respect to the rest of California. The faults of the Mojave and adjacent Death Valley have absorbed between 10 and 30 percent of the total slippage between the Pacific and North American plates over the last several million years.

Farther eastward, many parallel faults slice through the Basin and Range province. They absorb approximately 20 percent of the motion between the Pacific plate and

Figure 81 The Garlock Fault in the El Paso Mountains, Bernardino County, California. Courtesy of USGS

Figure 82 A view northeast across Death Valley to the Black and Funeral Mountains, Inyo County, California. Courtesy of USGS

the relatively stable North American continent east of the Great Basin. The Basin and Range province contains numerous fault block mountain ranges bounded by high-angle normal faults. The crust in this region is broken into hundreds of segments that were steeply tilted and raised nearly 1 mile above the basin, forming nearly parallel mountain ranges up to 50 miles long. Between the ranges are basins, where dry lake bed deposits are common because many of the low-lying areas once contained lakes. Great Salt Lake and the Bonneville Salt Flats in Utah are good examples. The region is literally being stretched apart due to the weakening of the crust by a series of downdropped blocks. As a result, 15 million years ago, the locations presently occupied by Reno, Nevada,and Salt Lake City, Utah, were 200 to 300 miles closer than they are today, moving apart at roughly 1 inch per year.

Death Valley, California (Figure 82), is the lowest spot on the North American continent. Although presently 280 feet below sea level, it had once been at an elevation of several thousand feet. The area collapsed when the continental crust thinned due to extensive block faulting in the region.

6

VOLCANIC RIFTS

All geologic activity taking place on the Earth's surface is an outward expression of the great heat engine in the interior of the planet. Continents separate and collide, mountains rise and trenches sink, and volcanoes erupt and faults quake. These actions shape the Earth and give it a unique character, both in the geologic sense and the biologic sense. The crust, the outermost layer of the Earth, is broken into several lithospheric plates (Figure 83) that are constantly interacting with each other, continuously changing and rearranging the face of the Earth.

Most of the crust is fashioned out of granitic and metamorphic rocks that constitute the bulk of the continents. The crust and the upper brittle mantle make up the lithosphere, which averages about 60 miles thick. The lithosphere rides freely on the semimolten outer layer of the mantle called the asthenoshpere. This structure is important for the operation of plate tectonics, and is found nowhere else in the Solar System.

CONTINENTAL RIFTING

Several times in the Earth's history, continents appear to have undergone cycles of collision and rifting (Figure 84). Smaller continental masses

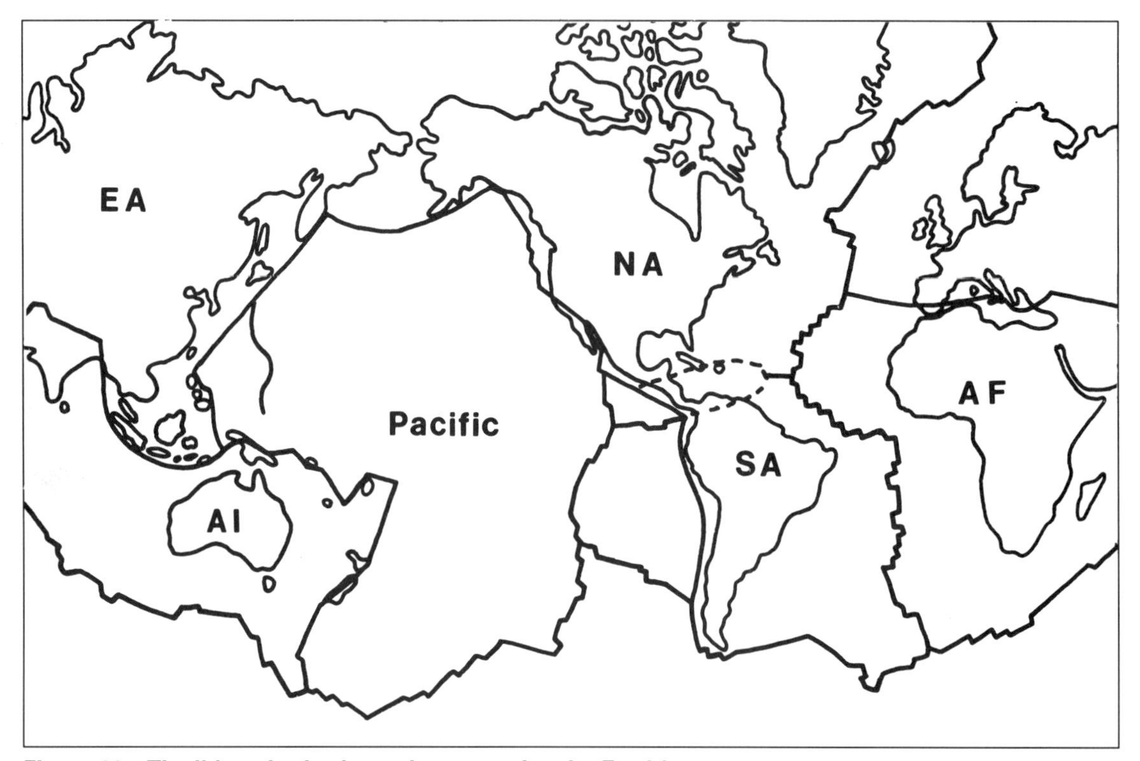

Figure 83 The lithospheric plates that comprise the Earth's crust.

collided and merged into larger continents. Millions of years later, the continents rifted apart, and the gaps filled with seawater to become new oceans. Periods of rifting occurred roughly 2.5 billion years ago. 2 billion years ago, 1.6 billion years ago, 1 billion years ago, 600 million years ago, and 170 million years ago.

Continents separate approximately 40 million years after rifting commences. They take about 160 million years to reach their greatest dispersal and to allow subduction of lithospheric plates in the newly formed oceans to begin. After the continents start to move back together, they might require another 160 million years to re-form a new supercontinent. The supercontinent might survive for about 80 million years before it again rifts apart. This cycle of rifting and patching repeats roughly every 450 million years.

The continents are driven by convection currents in the mantle generated by heat rising from the core and from the mantle itself. The latter is probably the greatest source of heat due to its great volume and large quantity of radiogenic isotopes. Rapid mantle convection leads to the breakup of supercontinents. This compresses the ocean basins, causing a rise in sea level and a transgression of the seas onto the land.

TABLE 7 DRIFTING OF THE CONTINENTS

Geologic division	Age (millions of years)	Gondwana	Laurasia
Quaternary period	3		Opening of Gulf of California
		Begin spreading near Galapagos Islands	
Pliocene epoch	11		Change spreading directions in eastern Pacific
		Opening of the Gulf of Aden	
			Birth of Iceland
Miocene epoch	25		
		Opening of Red Sea	
Oligocene epoch	40		
		Collision of India with Eurasia	Begin spreading in Arctic Basin
Eocene epoch	60		Separation of Greenland from Norway
		Separation of Australia from Antarctica	
Paleocene epoch	65		
			Opening of the Labrador Sea
		Separation of New Zealand from Antarctica	Opening of the Bay of Biscay

Geologic division	Age (millions of years)	Gondwana	Laurasia
		Separation of Africa from Madagascar and South America	
			Major rifting of North America from Eurasia
Cretaceous period	135		
		Separation of Africa from India, Australia, New Zealand and Antarctica	
Jurassic period	180		
			Begin separation of North America from Africa
Triassic period	230		
Permian period	280		

Convection currents arise not by the production of heat from the deeper core and mantle but by the conduction and loss of heat through the Earth's thin upper crust. Because continental crust is so much thicker than oceanic crust (Figure 85), it conducts heat less efficiently, making it function like an insulating blanket. If a supercontinent covers a portion of the Earth's surface where heat from the mantle accumulates under the crust, it causes the landmass to dome upward, creating a superswell. The bulging of the crust causes the supercontinent to rift apart, with individual broken off continents sliding off the superswell. The continents then move toward colder sinking regions in the mantle and become stranded over cool downflows of mantle rock.

When the continents are widely separated, heat from the mantle is more easily conducted through the newly formed ocean basins. After so much heat has escaped, the continents halt their outward progress and begin to return toward their places of origin. When today's continents have reached their maximum dispersal, perhaps millions of years from now, the crust of the Atlantic Ocean bordering the continents will grow sufficiently dense to sink into the mantle, thus creating subduction zones around the Atlantic Basin. The subduction of the oceanic crust into the mantle will begin the process of closing the Atlantic Ocean. Eventually, all continents will rejoin into a supercontinent over a large mantle downflow, and the cycle will begin anew.

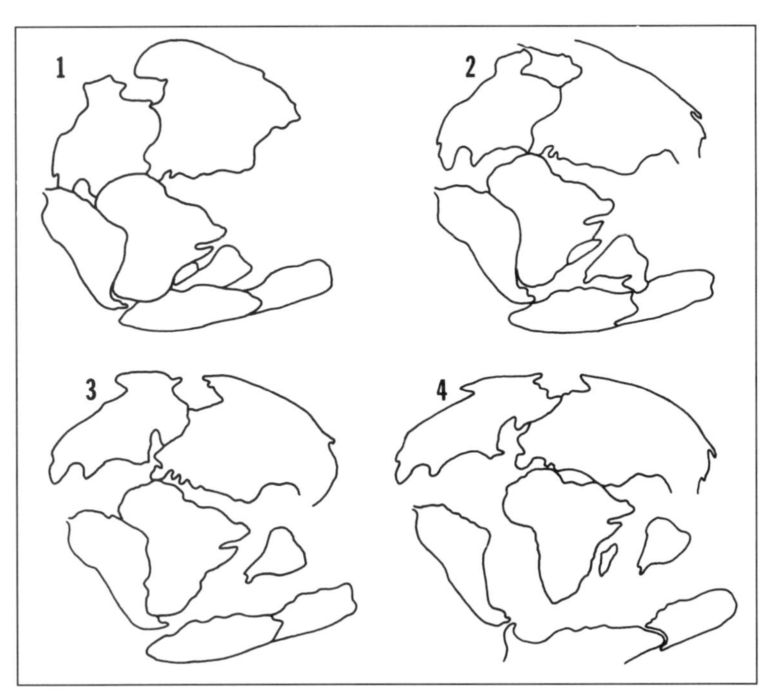

Figure 84 The breakup of the supercontinent Pangaea (1) 225, (2) 180, (3) 135, and (4) 65 million years ago.

The landmasses surrounding the Pacific Basin apparently have not undergone continental collision. The Pacific Ocean is a remnant of an ancient sea called the Panthalassa that has narrowed and widened in response to continental breakup, dispersal, and reconvergence in the area occupied by the present Atlantic Ocean. In this manner, several oceans have repeatedly opened and closed in the vicinity of the Atlantic Basin, while a single ocean has existed continuously in the vicinity of the Pacific Basin.

The Pacific plate, the largest tectonic plate, covering one quarter of the Earth's surface, was hardly larger than the United States following the last breakup of a supercontinent about 170 million years ago. The rest of the ocean floor consisted of several other plates that have since been subducted into the mantle to make room for the growing Pa-

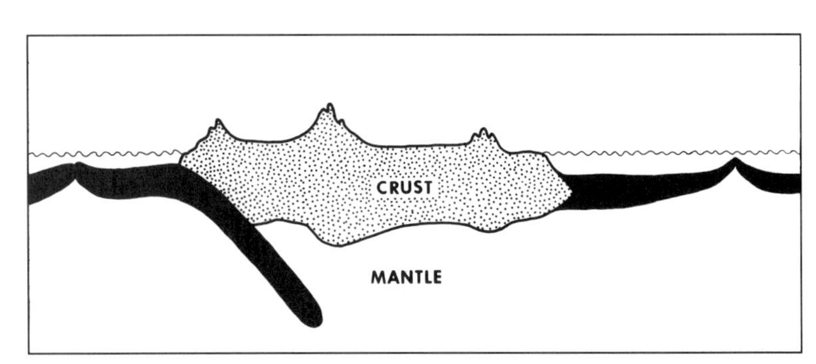

Figure 85 The Earth's crust is composed of continental granites and ocean basin basalts.

cific plate. The Pacific plate is presently being subducted under the continents surrounding it, whereas the oceanic crust in the Atlantic does not subduct beneath the surrounding continents.

CONTINENTAL RIFTS

The continental lithosphere, the solid outer shell of the Earth, is generally between 50 and 100 miles thick. Therefore, cracking open a continent would seem to be a formidable task due to its great thickness. Somehow, when continents rift into separate plates, thicker lithosphere must give way to thinner lithosphere. There is also a transition from a continental rift to an oceanic rift, accompanied by block faulting, whereby blocks of continental crust drop down along extensional faults where the crust is being pulled apart, resulting in a deep rift valley and a thinning of the crust.

The rifting of a continent begins with hot-spot volcanism at rift valleys. Hot spots act like geologic blow torches that burn holes through the crust and cause it to weaken. The hot spots are connected by rift valleys, along which the continent eventually splits apart. Mantle material rises up in giant plumes and underplates the crust with basaltic magma. This further weakens it, causing huge blocks to downdrop and form a series of grabens, or fault-bounded trenches.

The rising mantle convection currents spread out in opposite directions beneath the lithosphere and pull the thinning crust apart, forming a deep rift valley similar to the great East African rift system (Figure 86). In the process of rifting, large earthquakes strike the region as huge blocks of crust drop down along diverging faults. In addition, massive volcanic eruptions occur due to the abundance of molten magma rising up from the asthenosphere.

The crust underneath the rift is only 20 to 30 miles thick compared with 50 to 100 miles thick for the rest of the continent. As the crust continues to thin, magma chambers rise closer to the surface and volcanic eruptions become more prevalent. A marked increase in volcanism produces vast quantities of lava that flood onto the continental crust during the early stages of many rifts.

When the continent fragments, the rift valleys spread farther apart and eventually become a new ocean with an oceanic rift as seawater flows in and floods the region. As the rift continues to widen and deepen, it is replaced by a spreading ridge system, where hot mantle material wells up through the rifts to form new oceanic crust between the separated segments of continental crust. More than 2 million cubic miles of molten rock are released every time a supercontinent rifts over a hot plume, which explains the large increase in volcanic activity during the early stages of continental rifting.

Excellent evidence for the rifting of continents is found at the great East African Valley, which extends from the shores of Mozambique to the Red Sea, where it splits to form the Afar Triangle in Ethiopia. Afar is perhaps one of the best examples of a triple junction created by the doming of the crust over a hot spot. The Red Sea and the Gulf of Aden represent two arms of a three-armed rift, with the third arm heading into Ethiopia. When it finally breaks up, the continental rift will be replaced by an oceanic rift. This process is presently taking place in the Red Sea, which is rifting from north to south. The Gulf of Aden is a young oceanic rift between the ruptured continental blocks of Arabia and Africa, which have been diverging for more than 10 million years. The breakup of North America and Europe beginning about 170 million years ago might have been accomplished by the same upwelling of basaltic magma that is now occurring under the East African and Red Sea rifts.

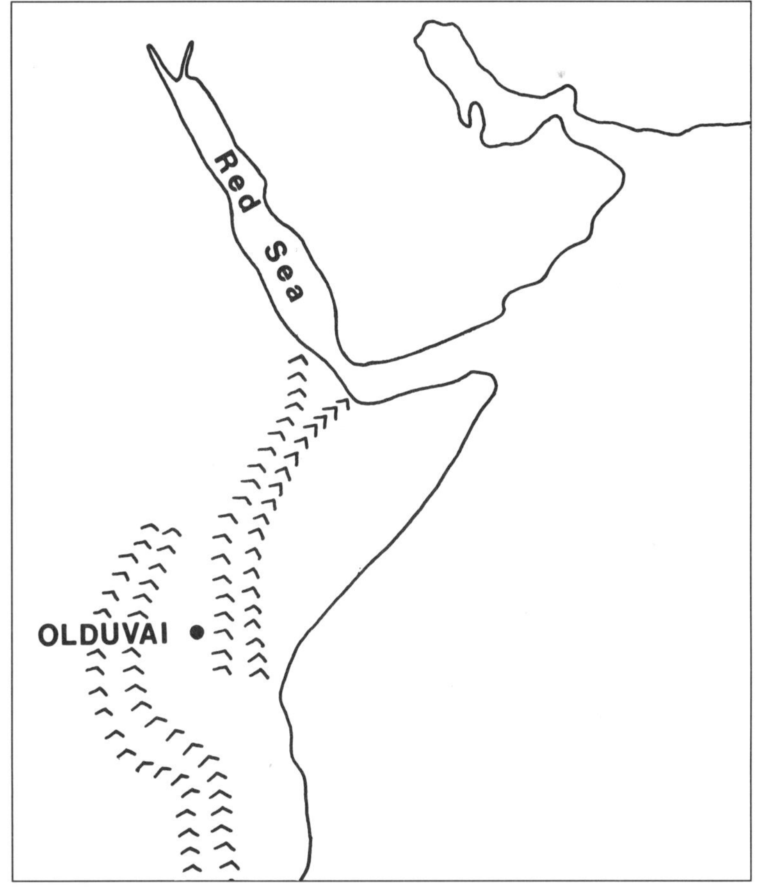

Figure 86 The East African rift system and its relation to one of the oldest sites of human habitation.

Extinct rift systems, where the spreading activity has ceased, or failed rifts, where a full-fledged spreading center did not develop, are sometimes overrun by continents. The western edge of North America has overridden the northern part of the now extinct Pacific rift system, creating the San Andreas Fault in California. A failed rift system beneath the central United States is responsible for the New Madrid Fault, which triggered three extremely large earthquakes in the winter of 1811–1812 that were felt several hundred miles away from the epicenter.

OCEANIC RIFTS

When the supercontinent Pangaea began to break apart about 170 million years ago, a rift developed in the present-day Caribbean. It sliced its way northward through the continental crust connecting North America, northwest Africa, and Eurasia and began to open up the Atlantic Ocean. The process took several million years along a zone that was several hundred miles wide. About 125 million years ago, the infant North Atlantic Ocean was already about 2.5 miles deep and had an active midocean ridge system making new oceanic crust. By about 80 million years ago, the North Atlantic became a full-fledged ocean (Figure 87).

About 125 million years ago, the South Atlantic began to develop, opening up from south to north. The rift propagated northward at a rate of several inches per year, comparable to the plate separation rate. The early stages of seafloor spreading in the Atlantic were similar to the processes now at work in the Red Sea that are separating Arabia from Africa. The entire process of opening the South Atlantic was completed in only about 5 million years.

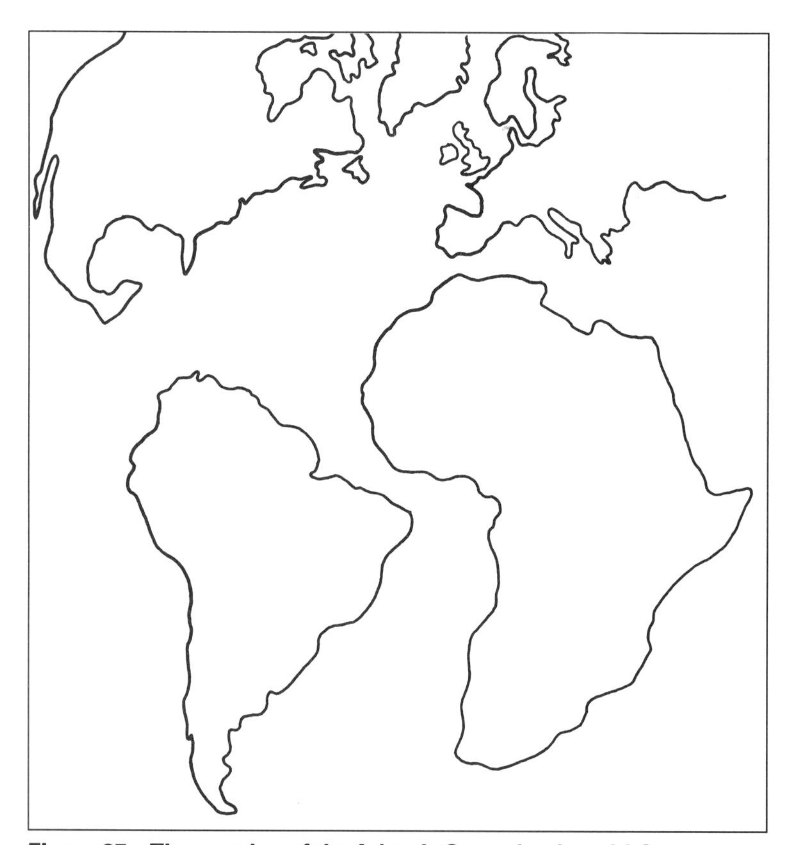

Figure 87 The opening of the Atlantic Ocean by the mid-Cretaceous.

The floor of the Atlantic acts like a huge conveyor belt that transports lithosphere outward from its point of origin at the Mid-Atlantic Ridge. The ocean floor at the crest of the ridge consists mostly of basalt. With increasing distance from the crest, the bare volcanic rock is covered with a thickening layer of sediments. As the plate cools, more material from the asthenosphere adheres to its underside, causing it to become thicker, heavier, and to sink deeper. This is why the ocean floor near the continental margins surrounding the Atlantic Basin is the deepest part of the Atlantic ocean (Figure 88).

The Mid-Atlantic Ridge bisects the Atlantic almost exactly down the middle, weaving halfway between continents and assuming the shapes of continental margins on opposite shorelines. It is the most extraordinary mountain range on Earth. The submerged mountains and undersea ridges form a continuous 45,000-mile-long chain (Figure 89) that is several hundred miles wide and up to 10,000 feet high. Although it is deeply submerged, the spreading ridge system is easily the most dominant feature on the planet, extending over an area larger than all major terrestrial mountain ranges combined. Running down the middle of the ridge crest is a deep trough like a giant

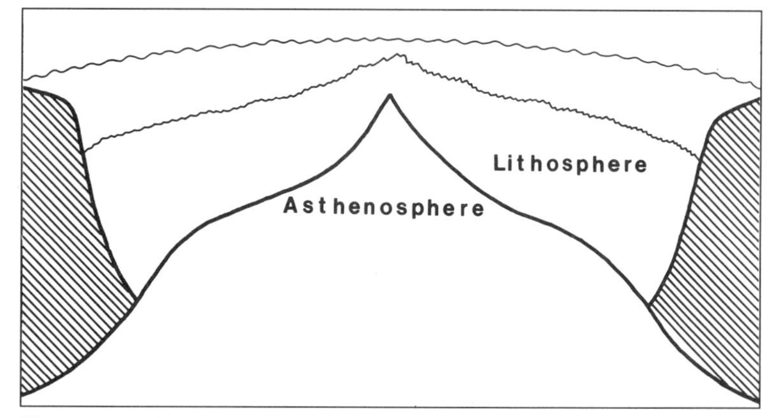

Figure 88 **The structure of the Mid-Atlantic Ridge.**

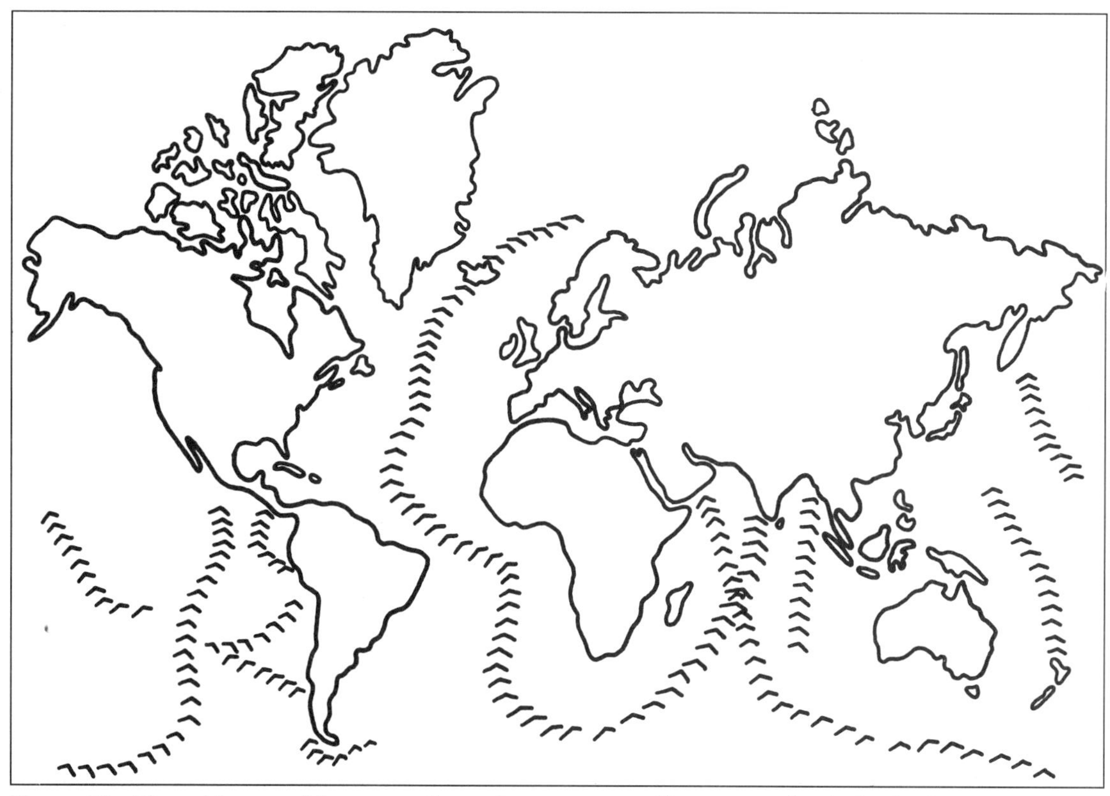

Figure 89 **The midocean ridges constitute the most extensive mountain chains in the world.**

crack in the ocean's crust. The trough is 4 miles deep in places, or four times deeper than the Grand Canyon, and up to 15 miles wide, which qualifies it as the largest canyon on Earth.

The axis of the Mid-Atlantic Ridge is offset laterally in a roughly east-west direction by transform faults, which are offsets consisting of a deep trough joining the tips of two segments of the ridge. Friction between segments gives rise to strong shearing forces, wrenching the ocean floor into steep canyons. Fracture zones offset the axis of the ridge, the largest of which is the Romanche Fracture Zone in the equatorial Atlantic. It extends for nearly 600 miles and has vertical relief of 4 miles. The fracture zone is flanked by several similar zones, producing a sequence of troughs and transverse ridges.

The ocean floor at the crest of the midocean ridge consists almost entirely of hard volcanic rock. The ridge system exhibits many unusual features, including massive peaks, sawtooth ridges, earthquake-fractured cliffs, deep valleys, and unusual lava formations (Figure 90). Along much of its length, the ridge system is carved down the middle by a sharp break or rift

Figure 90 Clusters of tube worms and sulfide deposits around hydrothermal vents near the Juan de Fuca Ridge on the East Pacific Rise. Courtesy of USGS

that is the center of an intense heat flow. In addition, the spreading ridges are the sites of frequent earthquakes and volcanic eruptions, as though the entire system were a series of giant cracks in the Earth's crust, from which molten magma bleeds out onto the ocean floor.

The magma that oozes onto the ocean floor at spreading ridges erupts as basaltic lava through long fissures in the trough between ridge crests and along lateral faults. The faults usually occur at the boundary between tectonic plates, where the brittle crust is pulled apart by the plate separation. Magma welling up along the entire length of the fissure forms large lava pools that harden to heal the crack.

The East Pacific Rise is a 6,000-mile-long rift system along the eastern edge of the Pacific plate that is the counterpart of the Mid-Atlantic Ridge and a member of the world's largest mountain chain. The rift system is a network of midocean ridges, most of which lies underwater. Each rift is a narrow fracture zone, where plates of the oceanic crust are being pulled apart by the action of plate tectonics.

Along the rift system the seafloor is spreading apart, causing molten rock to well up from the Earth's upper mantle and generate new sections of oceanic crust that move outward as though carried on a conveyor belt, driven by convection loops in the upper mantle. As the two newly separated plates move away from the rift, material from the asthenosphere adheres to their edges to form new oceanic lithosphere. This thickens the oceanic plate as it progresses away from the rift system.

On the crest of the East Pacific Rise, at the base of jagged basalt cliffs over 2 miles deep are lava flows and fields strewn with pillow lava. Active hydrothermal fields lie in areas where seawater that seeps downward near magma chambers is heated and expelled through hydrothermal vents. The undersea geysers build forests of exotic chimneys, called black smokers, that spew out hot water blackened with sulfur compounds (Figure 91).

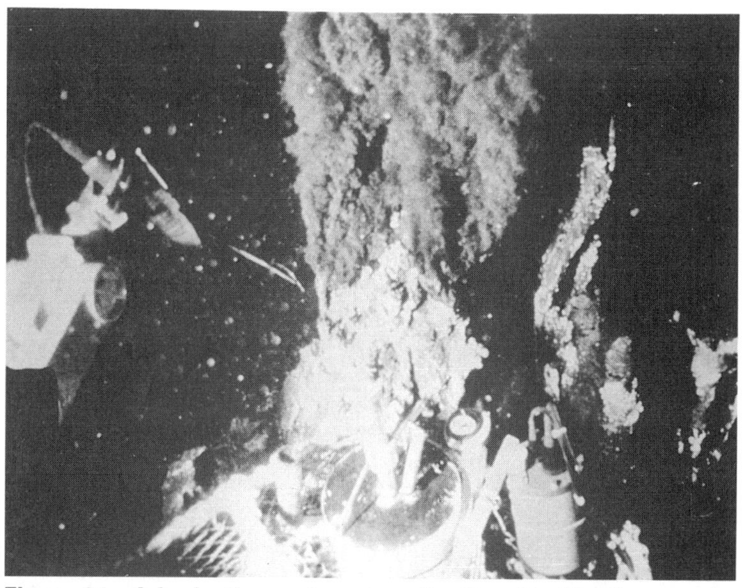

Figure 91 A hydrothermal vent on the East Pacific Rise pouring out black sulfide-laden hot water seen from the submersible *Alvin*, which is holding a temperature probe and a basket for samples. Photo by N. P. Edgar, courtesy of USGS

Magma contains significant amounts of sulfur. The circulating seawater below the ocean floor acquires sulfate ions and becomes acidic. This encourages the combination of sulfur with metals leached from the basalt. The sulfide minerals ejected from hydrothermal vents build up tall chimneys, some with branching vents. The black sulfide minerals drift along in the ocean currents and are deposited in depressions on the ocean floor to eventually become rich ore deposits.

The black smokers are also host to some of the Earth's most bizarre biology (Figure 92). Flourishing among the hydrothermal vents are perhaps the strangest animals ever encountered on Earth due to their unusual habitat. Large white clams up to 1 foot long nestle between black pillow lava. Giant white crabs scamper blindly across the volcanic terrain. Clusters of giant tube worms up to 10 feet tall sway in the ocean currents. They obtain their nutrition from bacteria that can metabolize sulfur compounds in the hydrothermal water, making their world totally independent of the sun for its energy, which comes instead from the Earth's interior.

Off the coast of Washington state, a field of seafloor geysers expels extremely hot brine, at temperatures between 350 and 400 degrees Celsius, into the near-freezing ocean water. Massive undersea volcanic eruptions from fissures on the ocean floor at spreading centers along the East Pacific Rise create large megaplumes of hot water. The megaplumes are produced by short periods of intense volcanic activity and are up to 50 or 60 miles wide. Apparently the ridge splits and spills out hot water at the same time the lava erupts in an act of catastrophic seafloor spreading. In a few hours, or at most a few days, over 100 million cubic yards of superheated water gushes from a large crack in the ocean crust up to several

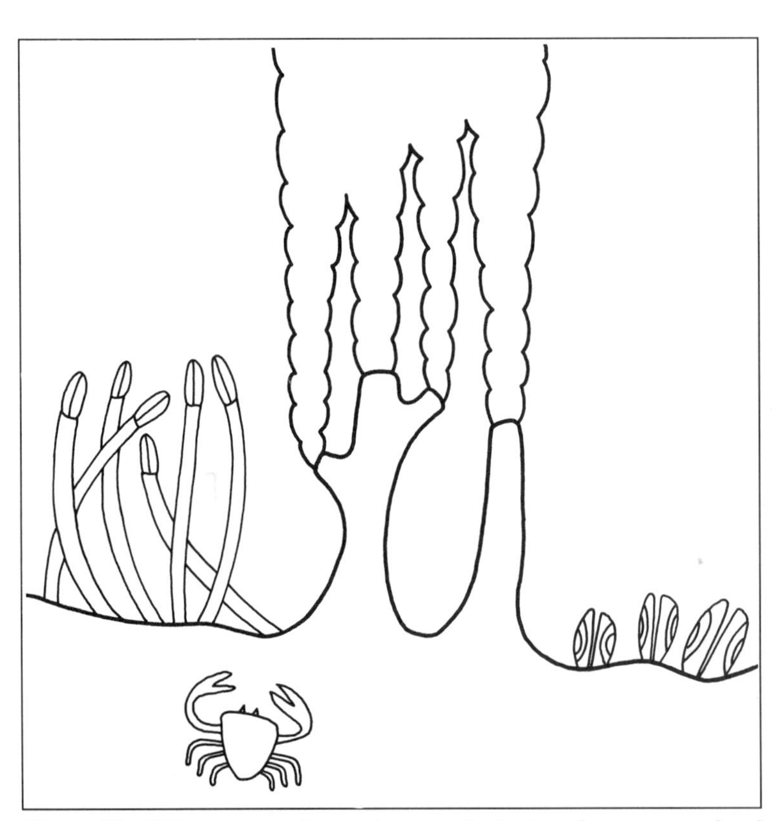

Figure 92 Tube worms, large clams and giant crabs are sustained by hydrothermal vents on the deep ocean floor.

miles long. When the seafloor splits in such a manner, it releases vast quantities of hot water that was held under great pressure beneath the surface, which might explain why the ocean continues to remain salty.

SEAFLOOR SPREADING

Seafloor spreading generates more than half of the Earth's crust. The process begins with hot rocks rising up by convection currents moving very slowly in the upper mantle. After reaching the underside of the lithosphere, the currents spread out laterally, cool, and descend back into the Earth's interior (Figure 93). The constant pressure against the bottom of the lithosphere produces fractures in the plate and forces it open.

As the convection currents flow out on either side of the fracture, they carry the separated parts of the lithosphere along with them, and the opening widens. This reduces the pressure, allowing the mantle rocks to melt and rise through the fracture zone. The molten rock passes through the lithosphere until it reaches the underside of the crust, where it forms magma chambers that supply material for new ocean floor.

The greater the supply of magma to the chamber, the higher the overlying ridge segment is elevated. The magma flows out from the trough between ridge crests, adding layer upon layer of basalt to both sides of the spreading ridge, creating new oceanic crust. The continents are carried passively along on the lithospheric plates created at spreading ridges and destroyed at subduction zones. Therefore, the engine that drives the birth and evolution of rifts and consequently the breakup of continents and the formation of ocean basins ultimately comes from the mantle.

New oceanic crust created at spreading ridges begins relatively thin and eventually thickens by the underplating of new lithosphere from the upper mantle and the accumulation of overlying sediment layers. By the time the lithosphere spreads out as wide as the Atlantic Ocean, the segment near continental margins, where the ocean is the deepest, is over 50 miles thick. Eventually, the lithosphere becomes so

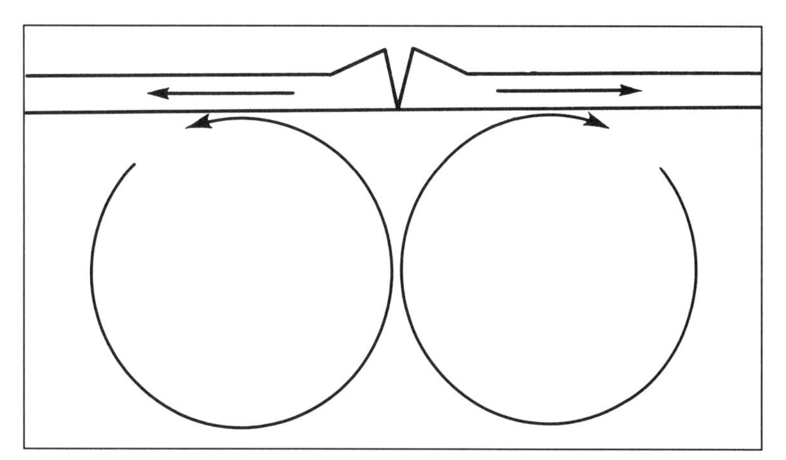

Figure 93 Convection currents in the mantle spread lithospheric plates apart.

thick and heavy it can no longer remain on the surface and subducts into the mantle, where it melts to provide material for new crust.

The mantle material below spreading centers, where new oceanic crust is created, consists mostly of peridotite, composed of iron and magnesium silicates. As the peridotite melts on its journey up to the base of the oceanic crust, a portion of it becomes highly fluid basalt. This is the most common magma erupted on the Earth's surface, and about 5 cubic miles of basaltic magma is added to the crust every year. Most of this volcanism occurs on the ocean floor at spreading centers, where the oceanic crust is being pulled apart.

Seamounts, underwater volcanoes associated with midocean ridges, can rise above sea level and become isolated volcanic islands. The volcanic islands of the East Pacific Rise are the Galapagos Islands west of Ecuador. Those of the Mid-Atlantic Ridge system include Iceland, the Azores, the Canary and Cape Verde Islands off West Africa, Ascension Island, and Tristan da Cunha. The St. Peter and St. Paul Rocks in the middle of the Atlantic just north of the equator are not volcanic but are fragments of the upper mantle that were uplifted close to the intersection of the St. Paul transform fault and the Mid-Atlantic Ridge.

TABLE 8 FLOOD BASALT VOLCANISM AND MASS EXTINCTIONS

Volcanic Episode	Million Years Ago	Extinction Event	Million Years Ago
Columbia River, USA	17	Low–mid Miocene	14
Ethiopia	35	Upper Eocene	36
Decca, India	65	Maastrichtian	65
		Cenomanian	91
Rajmahal, India	110	Aptian	110
South-West Africa	135	Tithonian	137
Antarctica	170	Bajocian	173
South Africa	190	Pliensbachian	191
E. North America	200	Rhaectian/ Norian	211
Siberia	250	Guadalupian	249

RIFT VOLCANOES

Over the past 250 million years, 11 episodes of massive flood basalt volcanism have occurred (Table 8). Great outpourings of basalt covered Washington, Oregon, and Idaho, creating the Columbia River Plateau (Figures 94 and 94A). Massive floods of lava poured onto an area of about 200,000 square miles and in places reached 10,000 feet thick. In only a matter of days, the volcanic eruptions spilled batches of basalt as large as 1,200 cubic miles, forming lava lakes up to 450 miles across.

These large eruptions created a series of overlapping lava flows that gave many exposures a terracelike appearance known as traps, from a Swedish word meaning "stairs." Many flood basalts are located near continental margins, where great rifts separated the present continents from a supercontinent 170 million years ago. The episodes of flood basalt volcanism were relatively short-lived events, with major phases generally lasting less than 3 million years.

Figure 94 Palouse Falls in Columbia River basalt, lower Snake River, Franklin-Whitman Counties, Washington. Photo by F. O. Jones, courtesy of USGS

Around 65 million years ago, a giant rift ran down the west side of India, and huge volumes of molten lava poured onto the surface. Nearly half a million square miles of lava up to 8,000 feet thick were released in less than half a million years. It blanketed much of west-central India, known as the Deccan Traps (Figure 95), in layers of basalt hundreds of feet thick. The rift

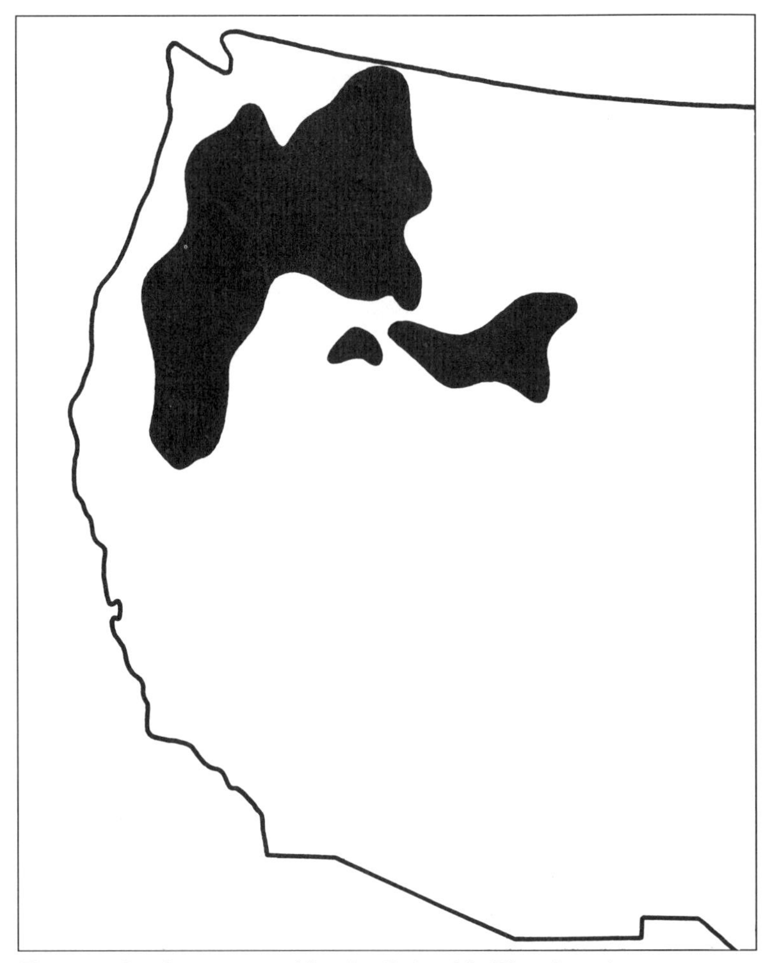

Figure 94A Area covered by the Columbia River basalts.

separated the Seychelles Bank from the mainland, leaving behind the Seychelles Islands as India continued to trek northward toward southern Asia, a journey it began about 150 million years ago, when it rifted away from the southern continent of Gondwana.

About 80 percent of oceanic volcanism occurs at spreading ridges (Table 9), where basaltic magma wells up from the mantle and spews onto the ocean floor. The spreading crustal plates grow by the steady accretion of solidifying magma. Over 1 square mile of new ocean crust, amounting to about 8 cubic miles of new basalt, is generated in this manner each year.

Iceland straddles the Mid-Atlantic Ridge, where the two plates that comprise the Atlantic Basin and adjacent continents are being pulled apart. The island is a broad volcanic plateau of the Mid-Atlantic Ridge that rises above sea level and is underlain by a large mantle plume, or hot spot. A steep-sided, V-shaped valley runs across the island from north to south. It is flanked by numerous active volcanoes, making Iceland one of the most volcanically active places on Earth (Figure 96).

For the past 25 to 30 million years, the Afar Triangle of the East African Rift Valley has been stewing with volcanism and has alternated between sea and dry land. The tiny African nation of Djibouti offers the unusual phenomenon of oceanic crust being extruded on dry land. The only other site in the world where seafloor spreading can be observed on land is Iceland.

The entire African rift is a complex system of tensional faults, indicating that the continent is in the initial stages of rupture. Much of the area has been uplifted thousands of feet by an expanding mass of molten magma lying just beneath the crust. This heat source is responsible for the numerous hot springs and volcanoes along the great rift valley. Some of the largest and oldest volcanoes in the world stand nearby, including Kenya and Mt. Kilimanjaro, at 19,590 feet the tallest mountain in Africa.

Seafloor spreading produces magma that slowly oozes out of the mantle and solidifies on the ocean floor. Occasionally, gigantic flows erupt with enough new basalt to pave the entire United States interstate highway system to a depth of 30 feet. Associ-

Figure 95 **Location of the Deccan Traps flood basalts in India.**

TABLE 9 GENERAL COMPARISON OF TYPES OF VOLCANISM

Characteristic	Subduction	Rift Zone	Hot Spot
Location	Deep-ocean trenches	Mid-ocean ridges	Interior of plates
Percent active volcanoes	80 percent	15 percent	5 percent
Topography	Mountains Island arcs	Submarine ridges	Mountains Geysers

Characteristic	Subduction	Rift Zone	Hot Spot
Examples	Andies Mts. Japan Is.	Azores Is. Iceland	Hawaiian Is. Yellowstone
Heat source	Plate friction	Convection currents	Upwelling from core
Magma temperature	Low	High	Low
Magma viscosity	High	Low	Low
Volatile content	High	Low	Low
Silica content	High	Low	Low
Type of eruption	Explosive	Effusive	Both
Volcanic products	Pyroclasts	Lava	Both
Rock type	Rhyolite Andesite	Basalt	Basalt
Type of cone	Composite	Cinder, Fissure	Cinder, Shield

ated with these huge bursts of basalt are megaplumes of warm, mineral-rich water. The mantle material that extrudes onto the surface is black basalt, which is rich in silicates of iron and magnesium. Most of the world's 600 active volcanoes are entirely or predominantly basaltic.

The magma from which the basalt is formed originated in a zone of partial melting in the Earth's upper mantle more than 60 miles below the surface. The semimolten rock at this depth is less dense than the surrounding mantle material and rises slowly toward the surface. As the magma ascends, the pressure decreases and more mantle material melts. The rising magma contributes to the formation of shallow reservoirs or feeder pipes that are the immediate source of volcanic activity.

The magma chambers closest to the surface exist under spreading ridges where the crust is only 6 miles or less thick. Large magma chambers lie under fast spreading ridges where the lithosphere is being created at a high rate, such as those in the Pacific. Narrow magma chambers lie under slow spreading ridges such as those in the Atlantic. As the magma chamber swells with magma and begins to expand, the crest of the spreading ridge is pushed upward by the buoyant forces generated by the molten rock.

Figure 96 Seawater being sprayed directly on a lava flow from the eruption of Heimaey to arrest it from infilling the harbor entrance at Vestmannaeyjar, Iceland, on May 4, 1973. Courtesy of USGS

The magma rises in narrow plumes that mushroom out along the spreading ridge. The plumes well up as a passive response to plate divergence, due to the release of pressure. Only the center of the plume is hot enough to rise all the way to the surface, however. If the entire plume were to erupt, it could build a massive volcano several miles high. Not all magma is extruded onto the ocean floor. Some solidifies within the conduits above the magma chamber and forms massive vertical sheets called dikes.

The two kinds of volcanism associated with midocean rift systems are fissure eruptions and those that build typical conical volcanic structures. During a fissure eruption, the magma oozes onto the ocean floor through long fissures in the trough between ridge crests and along lateral faults. The faults usually occur at the boundary between tectonic plates, where the brittle crust is split by the separation of the plates. Magma welling up along

Figure 97 Pillow lavas on the northwest shore of Ingot Island, Prince William Sound, Alaska. Photo by F.H. Moffit, courtesy of USGS

the entire length of the fissure forms large lava pools that solidify to seal off the rupture.

The two main types of lava formations associated with midocean ridges are sheet flows and pillow, or tube, flows. Sheet flows are more prevalent in the active volcanic zone of fast-spreading ridge segments like those of the East Pacific Rise, where in some places the plates are separating at a rate of 5 inches or more per year. Pillow lavas (Figure 97) erupt as though basalt was being squeezed out of a giant toothpaste tube. They are often found in slow spreading ridges, such as the Mid-Atlantic Ridge, where plates are separating at a rate of only about 1 inch per year and the lava is much more viscous. The manufacture of new oceanic crust in this manner explains why some of the most intriguing terrain features exist on the bottom of the ocean.

7

EARTHQUAKE FAULTS

When a continent is stretched by tectonic forces, originating deep within the Earth, massive blocks of crust bounded by faults collapse, creating what is called downdropped fault blocks, which further weaken the crust and set the stage for more faulting. These downdropped fault blocks are often associated with upraised sections of crust, providing a landscape of ridges and troughs. Not all faults are vertical, however, and many result from horizontal forces either produced by the compression of the crust or by two plates shearing past each other. The movement of the crust along faults causes earthquakes, which rupture the crust and rearrange the landscape.

FAULT BELTS

The vast majority of earthquakes are concentrated in a few narrow zones that wind around the globe (Figure 98). The area of greatest seismic activity lies on the boundaries between lithospheric plates, especially those associated with deep trenches and volcanic island arcs, where an oceanic plate is thrust under a continental plate. A continuous belt extends for thousands of miles through the world's oceans and coincides with midocean spreading ridges. The greatest amount of seismic energy is released along a path

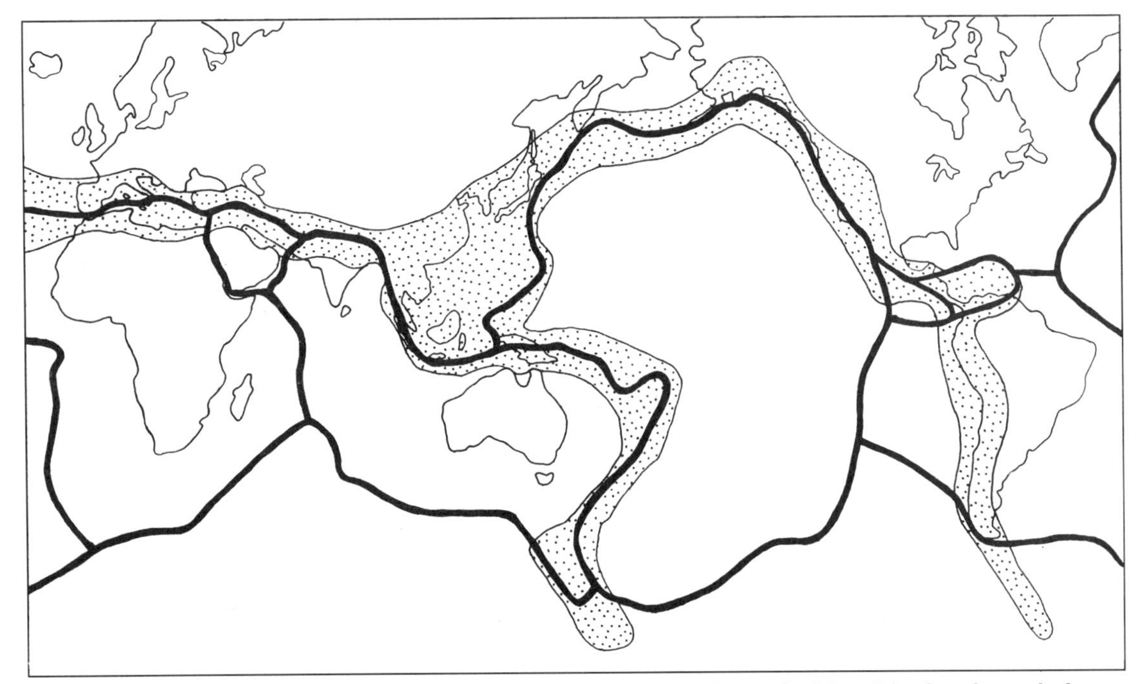

Figure 98 The majority of earthquakes occur in narrow belts that coincide with plate boundaries.

located near the outer edge of the Pacific Ocean, known as the circum-Pacific belt. This includes the San Andreas Fault, responsible for the numerous earthquakes that plague California.

The circum-Pacific belt follows the subduction zones that flank the Pacific Basin and corresponds to the "ring of fire," the rim of the Pacific Basin that contains most of the world's active volcanoes. In the western Pacific, the belt encompasses the volcanic island arcs that fringe the subduction zones, producing some of the largest earthquakes in the world (Figure 99). On the eastern side of the circum-Pacific belt, the Andes Mountain regions of Central and South America, especially in Chile and Peru, are known for some of the largest and most devastating earthquakes. In this century alone, nearly two dozen earthquakes of 7.5 magnitude or greater have taken place in Central and South America. The 1960 Chilean earthquake of 9.5 magnitude is the largest recorded anywhere in the world. The *magnitude scale*, equivalent to the Richter scale, is logarithmic; an increase of 1 magnitude signifies 10 times the ground motion and the release of 30 times the energy. (The 9.5 magnitude of the Chilean quake is actually a recomputed magnitude, measured on a somewhat different scale than the Richter scale.)

The entire western seaboard of South America is affected by an immense subduction zone just off the coast. The lithospheric plate on which the South America continent rides is forcing the Nazca plate to buckle under,

Figure 99 Rescue workers survey damages at the Christian College of the Philippines in the aftermath of an earthquake. Photo by Joseph Lancaster, courtesy of U.S. Navy

causing great tensions to build up deep within the crust. While some rocks are being forced deep down, others are pushed to the surface to raise the Andean mountain chain. The resulting forces are building great stresses into the entire region. When the strain grows too large, earthquakes roll across the coastal regions.

Another major seismic belt runs through the folded mountain belts that flank the Mediterranean Sea (Figure 100). The belt continues through Iran and past the Himalayan Mountains into China. At the eastern end of the Himalayan Range lies perhaps the most earthquake-prone region in the world. An enormous seismic belt, some 2,500 miles long, stretches across Tibet and much of China. Farther west, the Hindu Kush Range of north Afghanistan and the nearby Republic of Tadzhikistan experience many earthquakes. From there, the Persian arc spreads through the Pamir and Caucasus mountains and on to Turkey. The eastern end of the Mediterranean is a jumbled region of colliding plates, generating highly shaky ground.

Even in the so-called stable zones earthquakes occur, although less frequently than they do in the earthquake-prone regions. The stable zones are associated with shields composed of ancient rocks in the continental interiors. They include Scandinavia, Greenland, eastern Canada, parts of

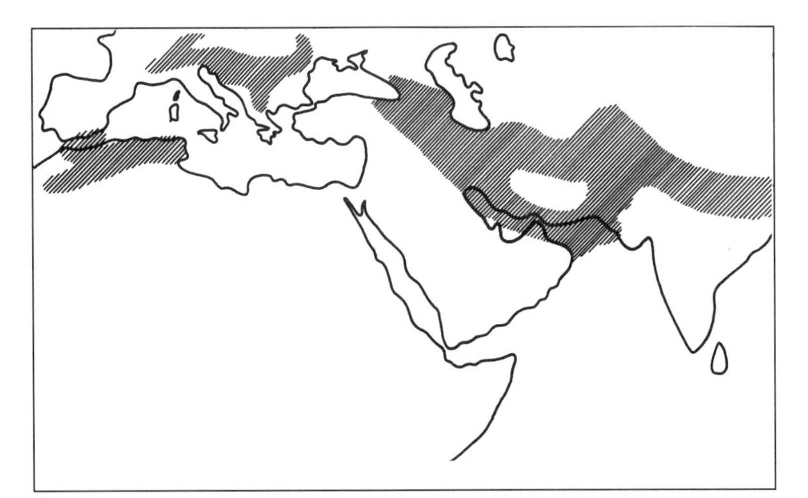

Figure 100 Active fold belts in Eurasia, resulting from the collisions of crustal plates.

northwestern Siberia and Russia, Arabia, the lower portions of the India subcontinent, the Indochina peninsula, almost all of South America except the Andes Mountain region, much of Australia, and the whole of Africa except the Great Rift Valley and northwestern Africa. When earthquakes occur in these regions, it is probably because the underlying crust has been weakened by previous volcanic or tectonic activity.

VERTICAL FAULTS

Not all movement along faults is horizontal. Vertical displacements that cause one side of the fault plane to be positioned higher relative to the other side are also common. Faults are classified according to the relationship between rocks on one side of the fault plane with respect to those on the other side (Figure 101). If the crust is pulled apart, one block, called the hanging-wall block, slides downward past the other block, called the footwall block, along a plane that is often steeply inclined. This is known as a gravity or normal fault, a historical misnomer because it was once thought to be how faults normally occurred.

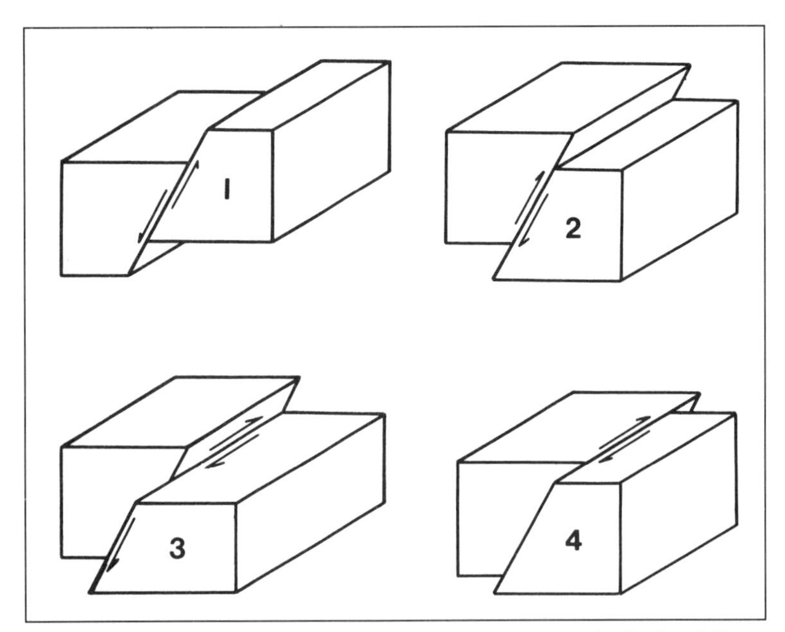

Figure 101 Fault types: 1. normal fault, 2. reverse fault, 3. oblique fault, 4. lateral fault.

Actually, most faults are produced by compressional forces, creating a reversed fault, so named because it is the reverse of a normal fault. In a reverse fault, one side of the fault plane is pushed above the other side along a vertical or steeply inclined plane. The 1964 Alaskan earthquake produced as much as 50 feet of vertical displacement, forming a high scarp along the fault zone. In the great 1960 Chilean earthquake, total displacements in areas immediately adjacent to the fault ranged upwards of 70 feet.

Some faults are neither all horizontal nor all vertical but consist of complicated diagonal movements. If a fault is a combination of both vertical and horizontal movements, it forms a complex fault system known as an oblique or scissors fault. The great Uinta Fault on the north side of the Uinta Mountains in northeast Utah (Figure 102) is an example of such a fault. The 1989 Loma Prieta earthquake in California ruptured a 25-mile-long segment of the San Andreas Fault. The faulting propagated upward along a dipping plane, resulting in a right oblique reverse fault. The earthquake raised the southwest side of the fault over 3 feet, contributing to the continued growth of the Santa Cruz mountains.

Figure 102 A view to the west toward the Uinta Fault from Bear Mountain, Uinta Mountains, Daggett County, Utah. Photo by W. R. Hansen, courtesy of USGS

Figure 103 The San Andreas Fault along Elkhorn scarp in the Carrizo Plains, California. Photo by R. E. Wallace, courtesy of USGS

LATERAL FAULTS

The San Andreas Fault in California (Figure 103) is a deep fracture zone that runs 650 miles northward from the Mexican border through southern California, plunging into the Pacific Ocean near the state's northern redwood forests. It forms the boundary between the North American and Pacific plates, separating southwestern California from the rest of the North American continent. The segment of California west of the San Andreas Fault along with the lithospheric plate on which it rides is sliding past the continental plate in a northwesterly direction at a rate of about 2 inches per year. The San Andreas Fault absorbs only about 75 percent of the slip. The rest is distributed over other faults in the region.

During the great 1906 San Francisco earthquake, a 260-mile section of the San Andreas Fault ruptured, and fences and roads crossing the fault were displaced several feet. The road between Point Reyes Station and Inverness was broken apart and offset horizontally by as much as 21 feet (Figure 104). During the fifty years before the earthquake, land surveys indicated displacements as much as 10 feet along the fault. Tectonic forces slowly deformed the crustal rocks on both sides of the fault, resulting in large lateral movements. Meanwhile, the rocks were bending and storing up elastic energy. Eventually, the forces holding the rocks together were overcome, and slippage occurred at the weakest point, generating one of the strongest earthquakes in California's recent history.

Not all segments of the San Andreas Fault have ruptured in historic times. Some of these faults are exposed on the surface, while others are buried deep below ground, where stresses along the San Andreas Fault increase with depth. A thrust fault that lies some 6 miles beneath the surface produced a damaging earthquake at Coalinga, California, in 1983. Thrust faults associated with the San Andreas Fault system might express themselves on the surface as a series of active folds. These thrust faults occur where a strike-slip fault ends, while the relative motion between blocks of crust continues, causing one block to push past the end of the fault and slide upward along a sloping plane. Meanwhile, the block of crust being pulled away might slip downward along a normal fault.

Faults exposed on the surface generally produce deep vertical fissures, created by the movement of crustal plates in different directions. But not all faults along which earthquakes occur are exposed, and most small earthquakes in California do not rupture the ground. Many of the earthquakes that are not associated with surface faults occur under folds, which are the product of successive earthquakes. Furthermore, folds can sometimes grow considerably during a large earthquake. For instance, an anti-

Figure 104 A road near Point Reyes Station is offset 20 feet by the San Andreas Fault during the 1906 San Francisco earthquake in California. Photo by G. K. Gilbert, courtesy of USGS

cline associated with the fault responsible for the great 1980 El Asnam earthquake in Algeria was uplifted more than 15 feet after the fault ruptured.

The relative motion of the Pacific and North American plates is right lateral, or dextral, because an observer on either side of the San Andreas Fault would notice the other block moving to the right. If the two plates slid smoothly past each other, earthquakes would not be so prevalent in the region. However, in the southern end of the fault and in the big bend area of the northern part of the fault, the plates tend to snag. When they attempt to break free, they generate earthquakes that are usually the largest in the state.

The 1906 San Francisco and the 1989 Loma Prieta earthquakes took place on a segment of the San Andreas Fault that runs through the Santa Cruz Mountains. The 1906 earthquake occurred on a vertical fault plane, where the two sides of the fault slid past each other horizontally, with very little vertical movement. Almost all the slippage along the San Andreas Fault is horizontal because the fault absorbs much of the motion between the Pacific and North American plates. The 1989 earthquake, however, occurred along a tilted surface that forced the southwest side of the fault to ride up over the northeast side.

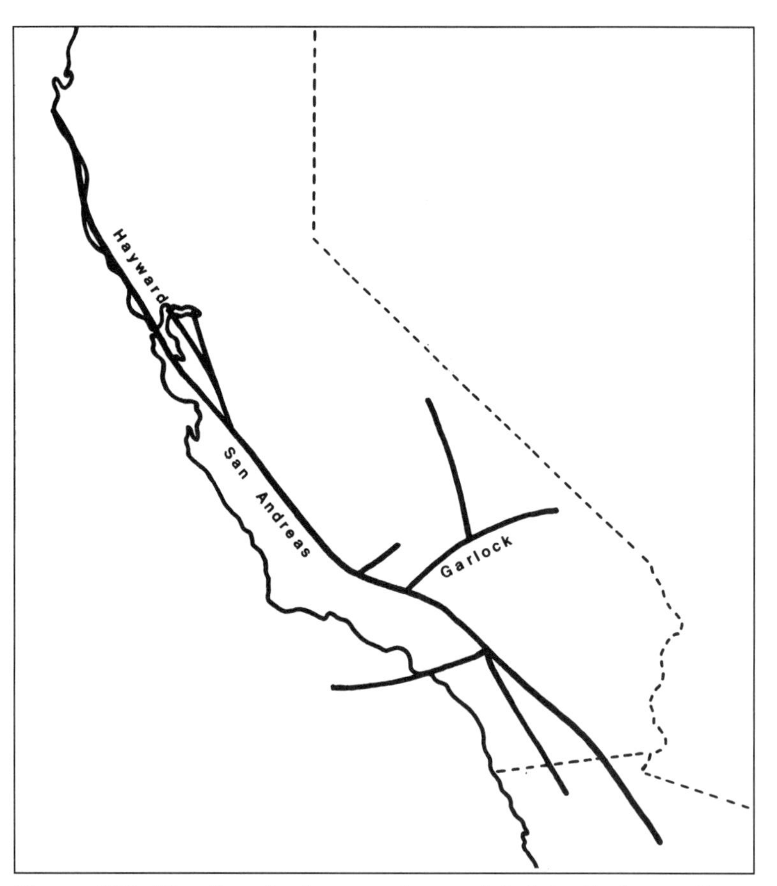

Figure 105 The San Andreas and associated faults in southern California.

Subsidiary faults of the San Andreas Fault system (Figure 105) include numerous parallel faults, such as the Hayward Fault that runs through suburban San Francisco, the Newport-Inglewood Fault, and numerous transverse faults. The Garlock Fault is a major east-trending fault. The faults of the Mojave Desert and adjacent Death Valley absorb about 10 percent of the slippage between the Pacific and North American plates. The complex crustal movements associated with these faults are responsible for most of the tectonic and geologic features of California, including the Sierra Nevada and Coast ranges.

A fault system that resembles the San Andreas is Scotland's Great Glen Fault. It dissects the country from coast to coast and caused the highlands to the north to slide past the lowlands to the south in a left-lateral, or sinistral, direction by as much as 60 miles since the late Paleozoic. The fault trace is marked by a belt of crushed and sheared rock up to 1 mile wide. A string of deep lakes runs along the fault, including Loch Ness, famous for its mythical monster.

Another fault system that mimics the San Andreas Fault is the 600-mile-long Red River Fault, running from Tibet to the South China Sea. When India collided with Asia some 45 million years ago, the fault allowed

Figure 106 An overthrust fault with Colorado Shale over Virgelle sandstone on the Missouri River near Virgelle, Chouteau County, Montana. Photo by I. J. Witkind, courtesy of USGS

Figure 107 The Blue Ridge escarpment, Appalachian Mountains, Macon County, North Carolina. Photo by A. Kieth, courtesy of USGS

Indochina to slide southeastward relative to south China in a process known as continental escape. As India continued to plow into Asia, it pushed Indochina at least 300 miles to the east, forcing it to jut out to sea, which rearranged the entire face of southeast Asia. The sideways escape of Indochina might have played a role in opening up a new ocean basin and creating the South China Sea.

Around 20 million years ago, the fault locked and halted the continental escape. This increased the stress on Asia, which thickened the crust and raised the Himalayan Mountains and the high Tibetan Plateau. Another extremely large strike-slip fault called the Altyn Tagh runs more than 1,200 miles along Tibet's northeast border. The fault has a high rate of slip, measuring over 1 inch per year. It is also allowing Tibet to escape to the east as India continues to plunge headlong into Asia at a rate of about 2 inches a year.

THRUST FAULTS

If a reverse fault plane is nearly flat and the movement is mainly horizontal for great distances, a thrust fault develops (Figure 106). Thrust faulting occurs when a highly compressed plate shears so that one section is lifted over another. Often, the slippage along a thrust fault is hidden deep below the surface. The overthrust belt from Canada to Arizona is a good example of this type of fault. Such faults are also responsible for trapping large reservoirs of oil, and geologists spend a great deal of time looking for these structures.

Horizontal thrusting can carry large volumes of crustal material over great distances. One way continents grow is by the emplacement of thin horizontal slices of material at the continental margins, which help create mountain ranges when two plates collide. For example, the Appalachians (Figure 107), extending some 2,000 miles from central Alabama to Newfoundland. were formed when North America, Eurasia, and Africa slammed into one another during the late Paleozoic.

The southern Appalachians are underlain by over 10 miles of sedimentary and metamorphic rocks that are essentially undeformed, whereas the surface rocks were highly deformed by the collision due primarily to thrust faulting. Like a rug thrown over a slippery floor, the sedimentary strata rode westward on top of the basement complex and folded over, buckling the crust into a series of ridges and valleys. Evidence of sedimentary layers under the core of the Appalachians suggests that thrusting involving basement rocks is responsible for the formation of mountain belts since the process of plate tectonics began over 2 billion years ago. Such thrusting and stacking of thrust sheets also might have been a major mechanism in the continued growth of the continents.

The counterpart of the Appalachians is the Mauritanide mountain range in western Africa. It is characterized by a series of belts running east to west that are similar in many respects to the Appalachian belts. The eastern parts of the range are composed of sedimentary strata partially covered by metamorphic rocks that have overridden the sediments from the west along thrust faults. Westward of this region are older metamorphic rocks that resemble those of the southern Appalachians. A coastal plain of younger horizontal rocks covers the rest of the region. In addition, before the opening of the Atlantic, a period of metamorphism and thrusting took place that was similar to what has happened to the Appalachians. Thus, the two mountain ranges are practically mirror images of each other.

When deep faults fail to break the surface during a major earthquake, the tremor might have been caused by a blind, low-angle thrust fault. The May 2, 1983, Coalinga, California, earthquake of 6.7 magnitude nearly leveled the town. The earthquake might have taken place on a thrust fault because there was no ground rupture, which should have occurred with any tremor greater than 6.0. The earthquake that struck Whittier, California, on October 1, 1987, was just under 6.0 in magnitude, and though the fault did not rupture the surface, the hills outside of town grew almost 2 inches.

Thrust faults can cause more damage than strike-slip faults for equal measures of magnitude. Strike-slip faults cause buildings to sway, and their flexible steel frames absorb most of the force. Thrust faults, on the other hand, suddenly raise and drop buildings inches at a time, creating tremendous forces that topple even the most well-designed structures.

HORSTS AND GRABENS

If a large block of crust is bounded by reverse faults and upraised with little or no tilting, it produces a long, ridgelike structure called a horst, from a German word meaning "ridge" (Figure 108). The Black Hills near Jerome

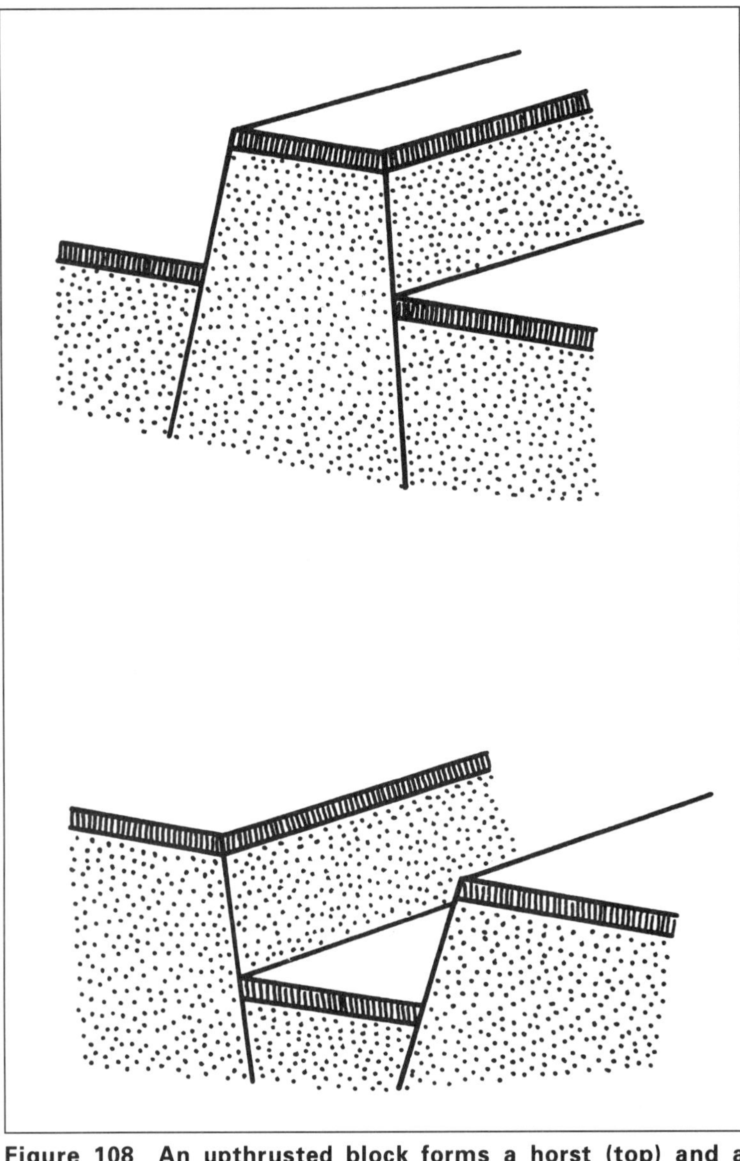

Figure 108 An upthrusted block forms a horst (top) and a downdropped block forms a graben (bottom).

in central Arizona are a horst with normal faults on the east and west. The hills comprise about 1,500 feet of flat-lying sedimentary strata overlying Precambrian granites. The horst is bounded on the east by the Verde Fault, which has a stratigraphic throw of about 1,500 feet. The horst is bounded on the west by the Cyote Fault, which dips steeply westward.

If a large block of crust is bounded by normal faults and downdropped, it produces a long, trenchlike structure called a graben, from a German word meaning "ditch." Grabens are generally much longer than they are wide. For example, the Rhine graben in Germany along the Rhine River Valley is 180 miles long and about 25 miles wide. Some grabens are expressed in the surface topography as linear structural depressions, with the flanking highland areas often consisting of horsts. Sometimes grabens are buried deep below ground, and the only way they are discovered is by drilling a series of boreholes across them.

Horsts and grabens are often found in association, forming long parallel mountain ranges and deep valleys like the great East African Rift, Germany's Rhine Valley, the Dead Sea Valley in Israel, the Baikal Rift in Russia, and the Rio Grande Rift in the American Southwest, which runs northward through central New Mexico into Colorado.

The rift valleys in Africa are a complex system of parallel horsts, grabens, and tilted fault blocks, with a net slip on the border faults of upwards of 8,000 feet. The eastern rift zone lies east of Lake Victoria, and extends for 3,000 miles from Mozambique to the Red Sea. The western rift zone lies west of Lake Victoria and extends northward for 1,000 miles. The rift just north of Lake Victoria filled with water to form Lake Tanganyika, the second deepest lake in the world. Lake Baikal at 6,000 feet deep has the distinction of being the world's deepest lake. It fills the Baikal rift zone, which is similar to the East African Rift.

The Basin and Range province of North America comprises numerous fault block mountain ranges that are bounded by high-angle normal faults. The crust in this region is broken into hundreds of pieces that have been tilted and upraised nearly a mile above the basin, forming about 20 nearly parallel mountain ranges 50 miles or more long. The region is literally being stretched apart due to the weakening of the crust by a series of downdropped blocks.

TRANSFORM FAULTS

Transform faults are created when pieces of oceanic crust slide past each other due to seafloor spreading (Figure 109). They were so named because they transform from active faults between spreading ridge axes to inactive fracture zones past the ridge axes. The ocean spreading ridge system does not form a continuous line, but is broken into small, straight sections called spreading centers. The movement of oceanic crust generated at the spreading centers produces a series of fracture zones, which are long, narrow linear regions up to 40 miles wide and consist of irregular ridges and valleys aligned in a step fashion.

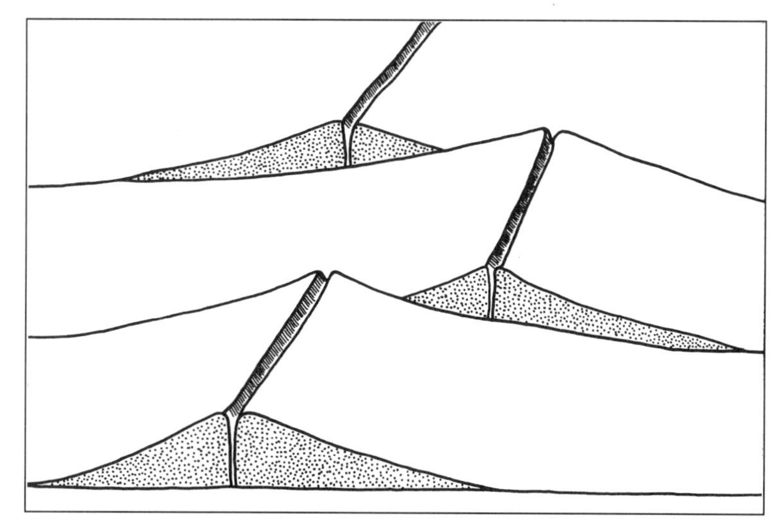

Figure 109 Transform faults at spreading centers on the ocean floor.

Transform faults are a few miles to a few hundred miles long and are encountered every 20 to 60 miles along the ridge system. The offsets are particularly common along the Mid-Atlantic Ridge, where the longer offsets consist of a deep trough joining the tips of two segments of the

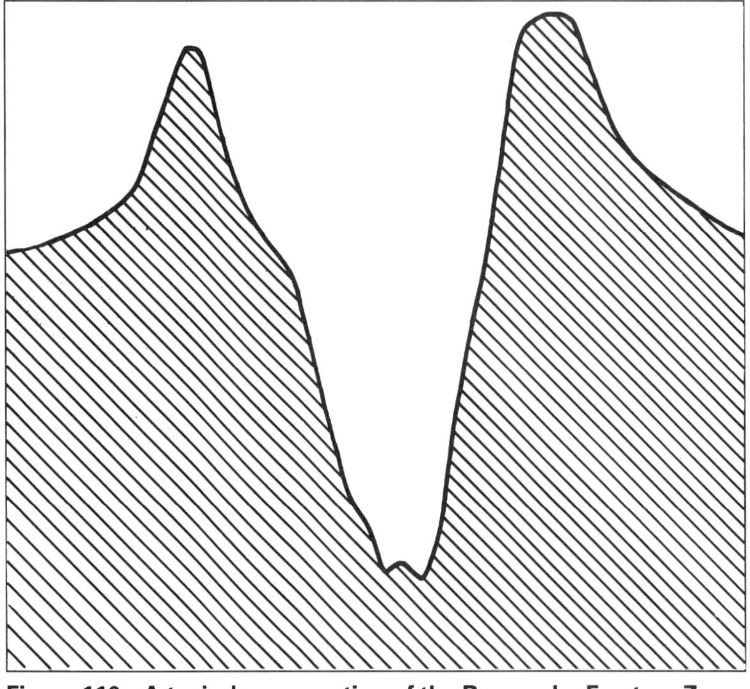
Figure 110 A typical cross section of the Romanche Fracture Zone.

ridge. Furthermore, several spreading centers that are 20 to 30 miles long are separated by nontransform offsets up to 15 miles wide. The end of one spreading center often runs past the end of another, and sometimes the tips of the segments bend toward each other.

The friction between lithospheric plates gives rise to strong shearing forces that wrench the ocean floor into steep canyons. The faults appear to result from the lateral strain created by movable plates on the surface of a sphere. This activity appears to be more intense in the Atlantic, where the Mid-Atlantic Ridge system is steeper and more jagged, than spreading ridges in the Pacific or Indian oceans.

In the equatorial Atlantic, a set of closely spaced fracture zones dissects the Mid-Atlantic Ridge. The largest of these zones is the Romanche Fracture Zone, which offsets the axis of the ridge by nearly 600 miles. The floor of the Romanche trench is as much as 5 miles below sea level, whereas the highest parts of the ridges on either side of the trench are less than 1 mile below sea level (Figure 110). This gives a vertical relief of four times that of the Grand Canyon. Many similar fracture zones that are almost equally impressive span the area, culminating in a sequence of troughs and transverse ridges that are several hundred miles wide. The resulting terrain is unmatched in size and ruggedness anywhere else on Earth.

The transform faults of the Mid-Atlantic Ridge are generally much rougher than those of the East Pacific Rise, which is the great spreading ridge system of the Pacific and counterpart to the Atlantic ridge system. Moreover, fewer widely spaced transform faults exist along the 6,000-mile length of the East Pacific Rise. The Rise marks the boundary between the Pacific and Cocos plates, which are separating at a rate of about 5 inches per year. Seafloor spreading is up to 10 times faster at the East Pacific Rise than it is at the Mid-Atlantic Ridge. Therefore, the crust affected by transform faults is younger, hotter, and less rigid in the Pacific than it is in the Atlantic, providing the undersea terrain with much less relief.

EARTHQUAKES

By far the most destructive natural forces in the world are earthquakes. The damage arising from a major earthquake is widespread, covering thousands

Figure 111 The Red Canyon Fault scarp developed during the August 1959 Montana earthquake, Gallatin County, Montana. Photo by J. R. Stacy, courtesy of USGS

of square miles. Not only do earthquakes destroy entire cities, but they completely change the landscape, often producing tall, steep-banked scarps (Figure 111) and causing massive landslides that scar the terrain. Successive earthquakes also upraise large folds and downdrop huge blocks of crust, forming ridges and valleys.

Every year, thousands of earthquakes occur, but fortunately only a few are powerful enough to be destructive (Table 10). During this century, the global average was about 18 major earthquakes of magnitude 7.0 or greater per year. During the 1980s, however, only 11 such shocks occurred on average per year. As for great earthquakes with magnitudes above 8.0, the century's average is 10 per decade, but the 1980s had only 4. Nevertheless, the number of large earthquakes appears to be on the rise.

The Earth's crust is constantly readjusting itself, as indicated by vertical and horizontal offsets on the surface. These movements are associated with large fractures in the Earth. The greatest earthquakes are produced by sudden slippage along major faults, sometimes with offsets of tens of feet taking place in seconds. Most faults are associated with plate boundaries, and most earthquakes are generated in zones where huge plates are shear-

TABLE 10 EARTHQUAKE MAGNITUDE SCALE AND EXPECTED INCIDENCE

Richter Scale	Earthquake Effects	Yearly Average
<2.0	Microearthquake—imperceptible	+600,000
2.0–2.9	Generally not felt but recorded	300,000
3.0–3.9	Felt by most people if nearby	50,000
4.0–4.9	Minor shock—damage slight and localized	6,000
5.0–5.9	Moderate shock—equivalent energy of atomic bomb	1,000
6.0–6.9	Large shock—possibly destructive in urban areas	120
7.0–7.9	Major earthquake—inflicts serious damage	14
8.0–8.9	Great earthquake—inflicts total destruction	10 a decade
9.0 and up	Largest earthquakes	1–2 a century

ing past or impinging upon each other. The interaction of plates causes rocks at their edges to strain and deform. The interaction occurs either near the surface, where major earthquakes occur, or several hundred miles below, where one plate is subducted under another.

The total slippage along a fault, accumulated from a number of earthquakes over a period of time, is used to estimate the velocity at which tectonic plates bounding the fault move past each other. By comparing this velocity with that computed by independent geologic, magnetic, and geodetic evidence, it is possible to determine how much of the plate's relative motion causes earthquakes and how much produces aseismic slip, which is ground movement along faults not associated with earthquakes.

Why seismic energy is released violently in some cases and not in others is not fully understood. In areas like Chile, noted for some of the world's largest earthquakes, all the motion between plates appears to be accompanied by earthquake slippage alone. The great 1960 Chilean earthquake, the largest in this century, occurred along a 600-mile-long rupture through the South Chile subduction zone. Generally, the longer the section of fault that breaks, the larger is the earthquake. Other processes that affect the magnitude of an earthquake include the speed at which the rupture travels over the fault, the frictional strength of the fault, and the drop in stress across the fault.

Earthquakes also occur in areas called stable zones but with much less frequency than those that occur at plate margins. The stable zones account for nearly two thirds of all continental crust. When earthquakes occur in these regions, they might be caused by the weakening of the crust due to compressive forces originating at plate boundaries. The crust might have been weakened by previous tectonic activity such as old faults and ancient mountain belts. Failed rift systems, where spreading centers did not fully develop, might be responsible for faults like the New Madrid in the central United States, which triggered three exceptionally large earthquakes in the winter of 1811–1812. Furthermore, the strong rock of plate interiors transmits seismic waves far more efficiently than the broken-up crust near plate boundaries. Therefore, earthquakes are felt over a much wider area.

Earthquakes also occur under folds, which do not rupture the Earth's surface like faults. Many of the world's major fold belts that raised mountain chains such as those bordering the Mediterranean Sea are earthquake-prone. During this century, large fold earthquakes have occurred in Japan, Argentina, New Zealand, Iran, and Pakistan. Most of these earthquakes appear to have taken place under young anticlines less than several million years old. These folds are the geologic product of successive earthquakes from compressional forces squeezing a continent.

8

GROUND FAILURES

Catastrophic ground failures are responsible for gouging out entire sections of the Earth's crust. Landslides on unstable slopes are a particular hazard in mountainous and hilly regions. Although not as hazardous as other geologic activity, landslides are more widespread and cause considerable damage. Rock falls are particularly spectacular, especially when involving large blocks that fall nearly vertically down a mountain face.

Under favorable conditions, the Earth can give way even on the gentlest slopes. The weakening of sediment layers due to earthquakes can cause massive subsidence. The Earth also moves when water is added to unstable sediments during heavy rains, causing various degrees of flowage. Submarine slides can be just as impressive as those on land and are responsible for much of the oceanic terrain along the outer margins of the continents.

CATASTROPHIC SLIDES

During the 1980 explosive eruption of Mount St. Helens, a wall of earth slid down the mountainside, creating one of the greatest landslides in modern times (Figure 112). It filled the valley below with debris covering an area of 20 square miles. One arm of the gigantic mass plowed through Spirit Lake at the base of the volcano and burst into the valley beyond, devastating everything in its path for 18 miles. Massive mudflows scoured the slopes of the volcano and jammed the Cowlitz and Columbia rivers all the way to the Pacific Ocean with debris and timber blown down by the tremendous blast.

Figure 112 Destruction from the 1980 eruption of Mount St. Helens, showing the mud-clogged Spirit Lake in the foreground. Photo by R. W. Emetaz, courtesy of USDA–Forest Service

One of the world's worst volcanic mudflows in recent history took place on November 13, 1985, when the eruption of Nevado del Ruiz Volcano in Columbia melted the mountain's icecap and sent floods and mudflows cascading 30 miles per hour down its sides. A 130-foot wall of mud and ash careened down the narrow canyon. When it reached the town of Armero 30 miles away, the mudflow spread out and flowed rapidly through the city streets, creating 10-foot-high waves. The deluge buried almost all of the town as well as other nearby villages and killed over 25,000 people.

In Italy, on the night of October 9, 1963, a torrent of water, mud, and rocks plunged down a narrow gorge, shot across the wide bed of the Piave River, and ran up the mountain slope on the opposite side, completely demolishing the town of Longarone and killing 2,000 of its residents. About

600 million tons of debris slid instantaneously into the reservoir below, which was only half-filled with water. The water was forced 800 feet above its previous level, and one great wave rose 300 feet above the dam and dropped into the gorge below. Constricted by the narrow gorge, the water increased speed tremendously and snatched up tons of mud and rocks as it raced on its destructive journey downstream.

One of the most spectacular examples of an avalanche occurred during the May 31, 1970, Peruvian earthquake that killed more than 18,000 people. A sliding mass of glacial ice and rock, 3,000 feet wide and about 1 mile long, rushed downslope. The avalanche traveled 9 miles to the town of Yungay and buried it under thousands of tons of rubble. It then shot across the valley and up the opposite bank where it partly destroyed the village of Huaraz (Figure 113). Flash flooding from broken mountain lake basins and from the avalanche-swollen waters of the Rio Santa River created a wave up to 45 feet high, compounding the earthquake's death and destruction.

Perhaps the most impressive rockfall in recent history occurred at Gohna, India, in 1893. A stupendous mass of rock loosened by the driving monsoon rains dropped 4,000 feet into one of the narrow Himalayan Mountain gorges. A great natural dam was formed 3,000 feet wide and 900 feet high and extended for 11,000 feet up and down the stream. The pile of broken rock, involving some 5 billion cubic yards, impounded a lake 770 feet deep. When the dam burst two years later, it caused a world record flood when about 10 billion cubic feet of water discharged in a matter of hours, with floodwaters cresting 240 feet high.

Figure 113 Destruction of adobe houses in Huaraz, Peru, from an avalanche created by the May 31, 1970, Peruvian earthquake. Courtesy of USGS

The most celebrated example of a rockfall in North America took place in Alberta, Canada, in 1903. A mass of strongly jointed limestone blocks at the crest of Turtle Mountain, which might have been undercut by coal mining carried on below the base, broke loose and plunged down the deep escarpment. Some 40 million cubic yards of material fell down the mountainside and washed through the small coal-mining town of Frank in one gigantic wave, killing 70 people along the way. The rockfall then swept up the opposite slope 400 feet above the valley floor.

LANDSLIDES

Landslides are a mass movement of soil and rock material downslope under the influence of gravity, caused primarily by earthquakes and severe weather. The main types of landslides are falls and topples, slides, and flows—either dry or wet. All slides result from the failure of earth materials under shear stress. They are initiated by an increase in shear stress and a reduction of shear strength along planes of weakness due primarily to the addition of water to a slope. The shear strength is determined by the slope geometry along with the composition, texture, and structure of the soil (Table 11). There might be changes in pore pressure and water content, which can reduce friction between rock layers.

Particles of rock, sand and snow dragged down a slope by gravity collide and rub against each other and the ground as they fall. With each such interaction, the particles change direction and lose energy to friction. Generally, the smaller the slope angle the less friction exists within the flow. The particles on the bottom in contact with the bed are slowed, while the rest of the particles glide over them in a tumbling, chaotic mass.

Most landslides in the United States occur in the Appalachian and Rocky Mountain regions and the coastal ranges along the Pacific Coast (Figure 114). Although individual landslides generally are not as spectacular as other violent forms of nature, they are more widespread and can cause major economic losses and casualties. The direct costs arising from damage to highways, buildings, and other facilities and indirect costs resulting from loss of productivity amount to more than $1 billion annually. Single large landslides can run up damage bills of millions of dollars. Fortunately, landslides have not resulted in a major loss of life as they have in other parts of the world, because most catastrophic slope failures in the United States generally take place in sparsely populated areas.

The majority of landslides occur during earthquakes. A landslide triggered by the 1959 Hebgen Lake, Montana, earthquake moved from north to south, gouging out a large scar in the mountainside (Figure 115). The debris traveled uphill on the south side of the valley and dammed the Madison River, creating a large lake. The 1971 San Fernando, California, earthquake unleashed nearly 1,000 landslides, distributed over a 100-square-mile area of remote and hilly mountainous terrain. During the 1976 Guatemala City earthquake, some 10,000 slides were triggered throughout an area of 6,000 square miles.

TABLE 11 SUMMARY OF SOIL TYPES

Climate	Temperate (humid; >160-in. rainfall)	Temperate (dry; <160-in. rainfall)	Tropical (heavy rainfall)	Arctic or Desert
Vegetation	Forest	Grass and brush	Grass and trees	Almost none, no humus develpment
Typical area	Eastern U.S.	Western U.S.		
Soil type	Pedalfer	Pedocal	Laterite	
Topsoil	Sandy, light colored; acid	Enriched in calcite; white color	Enriched in iron and aluminum; brick red color	No real soil forms because no organic material. Chemical weathering very low
Subsoil	Enrichd in aluminum, iron, and clay; brown color	Enriched in calcite; white color	All other elements removed by leaching	
Remarks	Extreme development in conifer forest; abundant humus makes ground-water acid. Soil light gray due to lack of iron	Caliche—name applied to accumulation of calcite	Apparently bacteria destroy humus, no acid available to remove iron	

In volcanic regions, seismic activity, uplift, and the existence of thick deposits of unconsolidated pyroclastic material create ideal conditions for landslides. The distribution of landslides is controlled by the seismic intensity, topographic amplification of the ground motion, lithology (rock type), slope steepness and regional fractures or other weaknesses in the rock. Heavy, sustained rainfall over a wide area can also trigger numerous landslides in volcanic terrain.

Landslides are also induced by the removal of lateral support by erosion from streams, glaciers, or waves, and longshore or tidal currents. They are also initiated by previous slope failures and human activity such as excavation. In addition, the ground can give way under excess loading by the weight of rain, hail, or snow.

Rockslides develop when a mass of bedrock is broken into many fragments during the fall and behaves like a fluid that spreads out in the valley below. It might even flow some distance uphill on the opposite side of the valley. Such slides are commonly called avalanches, but this term is generally applied to snowslides. Rockslides are usually large and destructive, involving millions of tons of rock. They are apt to develop if planes of weakness, such as bedding planes or jointing, are parallel to a slope, especially if the slope has been undercut by a river, glacier, or construction work.

Material that drops from a near vertical mountain face is called a rockfall or soilfall (Figure 116). Rockfalls can range in size from individual blocks plunging down a mountain slope to the failure of huge masses, weighing hundreds of thousands of

Figure 114 A rock/debris slide blocking Highway 1 south of Big Sur, California, from severe winter storms in 1982–1983. Photo by G.F. Wieczorek, courtesy of USGS

tons falling nearly straight down a mountain face. Individual blocks commonly come to rest in a loose pile of angular blocks at the base of a cliff, called talus. If large blocks of rock drop into a standing body of water, immensely destructive waves are set in motion. A 1958 earthquake in Alaska produced an enormous rockslide that fell into Lituya Bay and generated a wave of water surging 1,720 feet up the mountainside. Trees

Figure 115 The Madison Canyon slide from the August 1959 earthquake, Madison County, Montana.
Photo by J. R. Stacy, courtesy of USGS

were bowled over, and the shores along the bay were inundated with water that wiped out everything in its path (Figure 117).

Other forms of earth movements are slumps, which develop when a strong, resistant rock overlies weak rocks. Material slides down in a curved plane, tilting up the resistant unit, while the weaker rock flows out to form a heap. Unlike rockslides, slumps develop new cliffs just below those previous to the slump, setting the stage for a renewed slumping. Thus, slumping is a continuous process, and generally, many generations of slumps lie far in front of the present cliffs.

When loosened by rain or melting snow, ordinary soil on a steep hillside can suddenly turn into a wave of sediment, sweeping downward at speeds of over 30 miles per hour. Precipitation can free dirt and rocks by increasing the water pressure inside pores within the soil. As the water table rises and pore pressure increases, friction holding the top layer of soil to the hillside begins to drop until the pull of gravity overcomes it. Immediately before the soil begins to slide, the pore pressure drops, which signals that the soil is beginning to expand just before it starts to slide.

Soil slides occur in weakly cemented fine-grained materials that form steep stable slopes under normal conditions but fail during earthquakes. The size of the

Figure 116 Red Castle Peak in the Uinta Range, showing talus at the base, Summit County, Utah. Photo by W. R. Hansen, courtesy of USGS

Figure 117 Wave damage on the south shore of Lituya Bay, Lituya district, Alaska Gulf region. Photo by D. J. Miller, courtesy of USGS

area affected by earthquake-induced landslides depends on the magnitude and focal depth of the earthquake, the topography and geology of the ground near the fault, and the amplitude and duration of the tremor. Soil-flow failures caused by the great 1920 Kansu, China, earthquake killed an estimated 180,000 people. As the tremor rumbled through the region, immense slides rushed out of the hills, burying entire villages, damming streams, and turning valleys into instant lakes.

Soils and soft rocks that tend to swell or shrink due to changes in moisture content are called expansive soils. Damage to buildings and other structures built on expansive soils costs the nation as much as $7 billion annually. The soils are abundant in geologic formations in the Rocky Mountain region, the Basin and Range province, most of the Great Plains, much of the Gulf Plain, the lower Mississippi River Valley, and the Pacific

Coast. The parent materials for expansive soils are derived from volcanic and sedimentary rocks that decompose to form expansive clay minerals such as montmorillonite and bentonite. These materials are often used as drilling mud because of their ability to absorb large quantities of water. Unfortunately, this characteristic also causes them to form highly unstable slopes.

LIQUEFACTION

Ground failures during earthquakes and violent volcanic eruptions, resulting from the failure of subterranean sediments that are saturated with water, are caused by liquefaction. Earthquakes can turn a solid, water-saturated bed of sand underlying less-permeable surface layers into a pool of pressurized liquid that seeks a way to the surface, sometimes causing localized flooding. Generally, the younger and looser the sediment, and the closer the water table is to the surface, the more susceptible the soil is to liquefaction.

Liquefaction causes clay-free soils, primarily sands and silts, to temporarily lose strength and behave like viscous fluids rather than as solid materials. It occurs when seismic shear waves pass through a saturated granular soil layer. This distorts its structure and causes some of the void spaces to collapse in loosely packed sediments. Each collapse transfers stress to the pore water surrounding the grains. Disruptions to the soil generated by these col-

Figure 118 Ground fracture in Forest Acres area from lateral spreading during the March 27, 1964, Alaskan earthquake, Seward district, Alaska Gulf region. Photo by R. D. Miller, courtesy of USGS

lapses increases pressure in the pore water, causing drainage to occur. If the drainage is restricted, the pore-water pressure builds up. When the pore-water pressure reaches the pressure exerted by the weight of the overlying soil, grain contact stress is temporarily lost and the granular soil layer flows like a fluid.

The three types of ground failure associated with liquefaction are lateral spreads, flow failures, and loss of bearing strength. Lateral spreads are the lateral movement of large blocks of soil due to liquefaction in a subsurface layer caused by earthquakes (Figure 118). They generally develop on gentle slopes of less than 6 percent. Horizontal movements on lateral spreads can extend up to 15 feet, but where slopes are particularly favorable and the duration of the tremor is long, lateral movement might extend up to 10 times farther. Lateral spreads usually break up internally, forming numerous fissures and scarps.

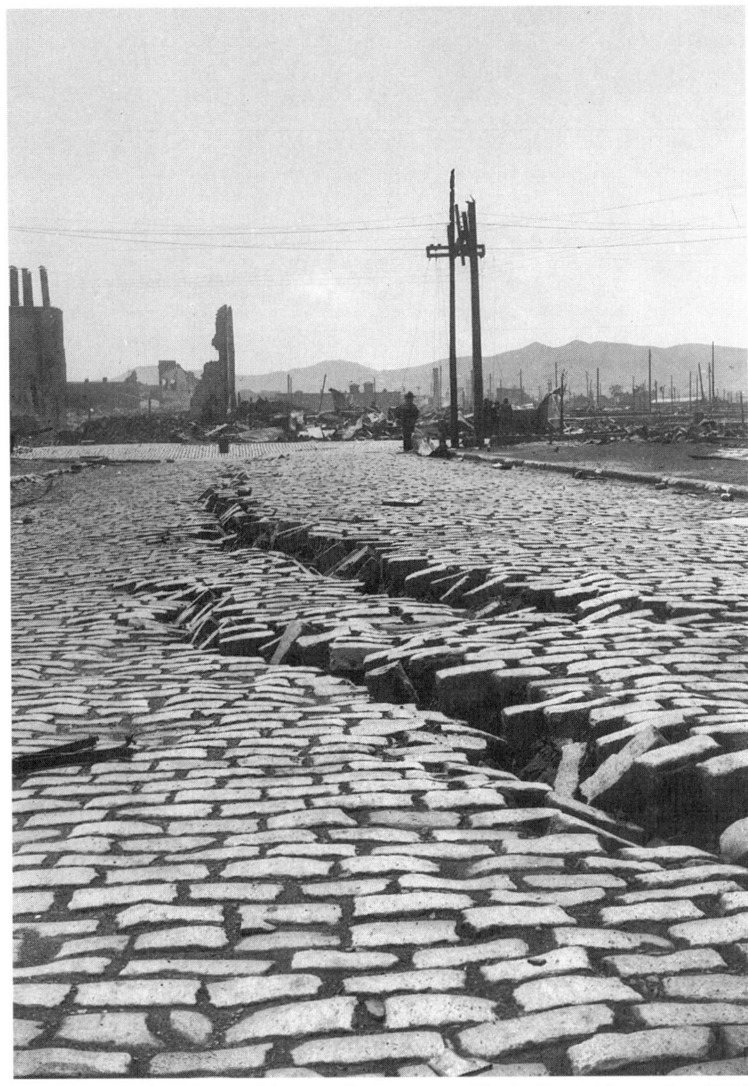

Figure 119 Cracks on Bluxom Street near Sixth Street from the 1906 San Francisco earthquake, San Francisco County, California. Photo by G. K. Gilbert, courtesy of USGS

During the 1964 Alaskan earthquake, more than 200 bridges were damaged or destroyed by lateral spreading of floodplain deposits near river channels. These spreading deposits compressed bridges over the channels, buckled decks, thrust sedimentary beds over abutments and shifted and tilted abutments and piers. Lateral spreads are also destructive to pipelines. During the 1906 San Francisco earthquake, a number of major watermain breaks occurred, which hampered firefighting efforts. The inconspicuous ground failure displacements, some as large as 7 feet, were largely responsible for the destruction of San Francisco (Figure 119).

Flow failures are the most catastrophic type of ground failure associated with liquefaction. They consist of liquefied soil or blocks of intact material riding on a layer of liquefied soil. Flow failures usually move dozens of feet, but under certain geographical conditions they can travel several miles at speeds of up to many miles per hour. They usually form in loose saturated sands or silts on slopes greater than 6 percent and originate both

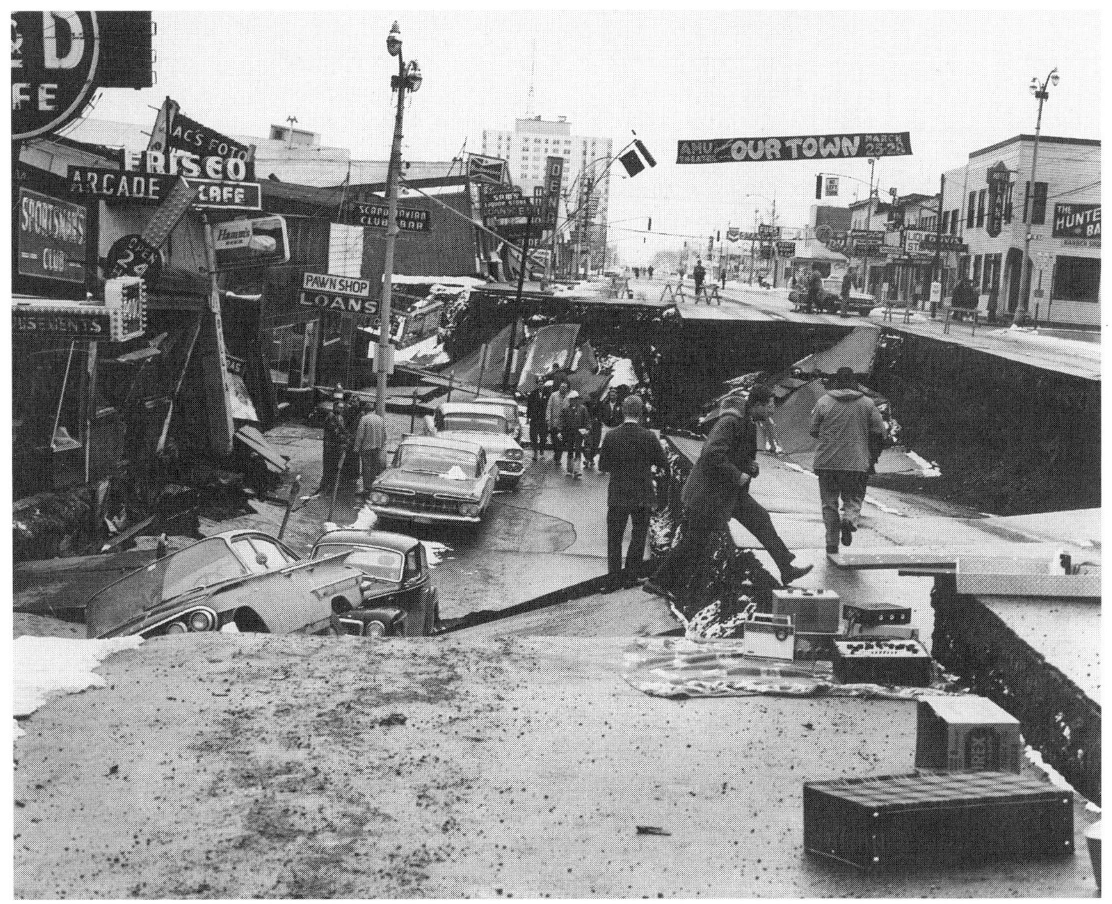

Figure 120 The collapse of Fourth Avenue in Anchorage from the March 27, 1964, Alaskan earthquake, Anchorage district, Alaska. Courtesy of USGS and U.S. Army

on land and on the seafloor. Flow failures on land can be particularly catastrophic. The 1920 Kansu, China, earthquake induced several flow failures as much as one mile in length and breadth that killed as many as 180,000 people.

Most clays lose strength when disturbed by earthquakes, and if the loss of strength is large some clays, called quick clays, might fail. Quick clay is composed primarily of flakes of clay minerals arranged in very fine layers, with a water content that often exceeds 50 percent. Ordinarily, quick clay is a solid capable of supporting over a ton per square foot of surface area. However, the slightest jarring motion from an earthquake can immediately turn it into a liquid.

During the 1964 Alaskan earthquake, landslides and ground subsidence caused the greatest damage to man-made structures. The ground beneath Valdez and Seward literally gave way, and both waterfronts floated toward the sea. In Anchorage, houses were destroyed when 200 acres were carried toward the ocean. The five large landslides that affected parts of Anchorage are examples of spectacular failures of clays sensitive to ground motions (Figure 120). The slides resulted from the failure of layers of quick clay along with other layers composed of saturated sand and silt. Because of the severity of the earthquake, loss of strength occurred in the clay layers and liquefaction occurred in the sand and silt layers. These were the major contributing factors to the landsliding that destroyed a major portion of the city.

MASS WASTING

Not all earth movements are induced by earthquakes. Many are caused by mass wasting, which is the mass transfer of material downslope by the direct influence of gravity. Mass wasting causes slipping, sliding, and creeping even down the gentlest slopes. Creep is the slow downslope movement of bedrock and overburden, the soil overlying bedrock. It is recognized by downhill tilted poles and fence posts, indicating a more rapid movement of near-surface soil material than that below.

Normally, trees are unable to root themselves, and only grass and shrubs grow on the slope. In some cases where the creep is slow, the trunks of trees are bent, and after the trees are tilted, new growth attempts to straighten them. But if the creep is continuous, the trees lean downhill in their lower parts and become progressively straighter higher up. Creep might be very rapid where frost action is prominent. After a freeze-thaw sequence, material moves downslope due to the expansion and contraction of the ground.

A rise in the water content of the overburden increases the weight and reduces stability by lowering resistance to shear, resulting in an earthflow, which is a more visible form of movement. Earthflows are characterized by grass-covered, soil-blanketed hills. Although generally they are minor features, some can be quite large and cover several acres. Earthflows usually have a spoon-shaped sliding surface, whereupon a tongue of overburden breaks away and flows for a short distance. An earthflow differs from creep in that a distinct, curved scarp is formed at the breakaway point.

Figure 121 The Slumgullion mudflow, Himsdale County, Colorado, in 1905. Photo by W. Cross, courtesy of USGS

With a further increase in water content, an earthflow might grade into a mudflow (Figure 121). The behavior of mudflows is similar to that of a viscous fluid, often carrying a tumbling mass of rocks and large boulders. They are also produced by rain falling on loose pyroclastic material (rocks formed by volcanic explosion) on the flanks of certain types of inactive volcanoes. Mudflows are the most impressive feature of many of the world's deserts. Heavy runoff forms rapidly moving sheets of water that pick up huge quantities of loose material. The floodwaters flow into the main stream, where all the muddy material is suddenly concentrated in the main channel. The dry stream bed is rapidly transformed into a flash flood that moves swiftly downhill, in some cases with a steep, wall-like front.

This type of mudflow could cause considerable damage as it flows out of mountain ranges. Eventually, the loss of water by percolation into the ground thickens the mudflow until it ceases to flow. Mudflows can carry large blocks and boulders onto the floor of the desert basins far beyond the base of the bordering mountain range. Often, huge monoliths that were rafted out beyond the mountains by swift-flowing mudflows are left standing in the middle of nowhere.

Mudflows arising from volcanic eruptions are called lahars, from the Indonesian word meaning "mudflow," due to their large occurrence in this region. Lahars are masses of water-saturated rock debris that move down steep slopes of volcanoes in a manner resembling the flowage of wet concrete. The debris is commonly derived from masses of loose unstable rock deposited on the volcano by explosive eruptions. The water is provided by rain, melting snow, a crater lake, or a reservoir adjacent to the volcano. Lahars can also be initiated by a pyroclastic or lava flow moving across a snowfield, causing it to rapidly melt. They can be either cold or hot, depending on whether hot rock debris is present.

The speed of the lahars depends mostly on their fluidity and the slope of the terrain. They can move swiftly down valley floors for up to 50 miles or more at speeds exceeding 20 miles per hour. Lava flows that extend into areas of snow or glacial ice might melt them, producing floods as well as lahars (Figure 122). Flood-hazard zones extend considerable distances down some valleys. For the volcanoes in the western Cascade Range these zones can reach as far as the Pacific Ocean.

The most common triggering mechanisms for mass wasting include vibrations from earthquakes or explosions that break the bond holding the slope together, overloading the slope so it can no longer support its new weight, undercutting at the base of the slope, and oversaturation with water. The effect of water is twofold. It adds to the weight of the slope and

Figure 122 The May 18, 1980, eruption of Mount St. Helens caused extensive flooding and sedimentation along the Cowlitz River, Cowlitz County, Washington. Courtesy of USGS

lessens the internal cohesion of the overburden. Although the effect of water as a lubricant is commonly considered to be its main role, this is actually quite limited. The main effect is the loss of the cohesion of the material by filling the spaces between soil grains with water.

Another type of movement of soil material is called frost heaving. It is associated with cycles of freezing and thawing mainly in the temperate climates. Frost heaving thrusts boulders upward through the soil, both by a pull from above and by a push from below. If the top of the rock freezes first, it is pulled upward by the expanding frozen soil. When the soil thaws, sediment gathers below the rock, and it settles at a slightly higher level. The expanding frozen soil lying below also heaves the rock upward. After several frost-thaw cycles, the boulder finally comes to rest on the surface, which can be a major annoyance to northern farmers. Rocks have also been known to push through highway pavement, and fence posts have been known to be shoved completely out of the ground.

Frost action can produce mechanical weathering by exerting pressures against the sides of cracks and crevices in rocks when water freezes inside them, resulting in frost wedging. This widens the cracks, while surface weathering tends to round off the edges and corners, providing a landscape resembling multitudes of miniature canyons up to several feet wide that are carved into solid bedrock.

The Arctic has another bizarre manifestation of frost heaving, consisting of regular polygonal-patterned ground that has the appearance of a giant tiled floor (Figure 123). In many parts of the tundra, soil and rocks are fashioned into orderly patterns that are found in most of the northern lands and alpine regions, where the soil is exposed to moisture and seasonal freezing and thawing. The polygons range in size from a few inches across, when composed of small pebbles, to several tens of feet wide when large boulders form protective rings around mounds of soil (Figure 124).

Figure 123 Patterned ground on the northern Alaskan seacoast near Barrow, Barrow district, northern Alaska. Photo by L. T. Pewe, courtesy of USGS

The regular polygonal patterned ground in the Arctic regions is formed by the movement of soil of mixed composition up-

Figure 124 Sorted stone circles caused by frost action in Alaska Range, south-central Alaska. Photo by T. L. Pewe, courtesy of USGS

ward toward the center of the mound and downward under the boulders, making the soil move in convective cells. The coarser material composed of gravel and boulders is gradually shoved radially outward from the central area, leaving the finer materials behind in a ring shape.

SUBMARINE SLIDES

Many of the largest and most damaging flow failures have taken place undersea in coastal areas. Submarine flow failures carried away large sections of the port facilities at Seward, Whittier, and Valdez, Alaska, during the 1964 earthquake (Figure 125). Flow failures can generate large sea waves, or tsunamis, that overrun parts of the coast. Coastal landslides are produced when a sea cliff is undercut by wave action and falls into the ocean. Excessive rainfall along the coast can also lubricate sediments, causing huge blocks to slide into the sea. Submarine slides that move down steep continental slopes have been known to bury undersea telephone cables beneath thick layers of sediment.

Figure 125 A railroad yard and warehouse damaged at Seward due to submarine slides from the March 27, 1964, Alaskan earthquake, Seward district, Alaska Gulf region. Courtesy of USGS

Submarine slides are also responsible for carving out deep submarine canyons in continental slopes. The slides consist of sediment-laden water, and because they are heavier than the surrounding seawater they move swiftly along the ocean floor, eroding the soft bottom material (Figure 126). These muddy waters are called turbidity currents. They can move down the gentlest slopes and transport fairly large rocks. Turbidity currents can also be initiated by river discharge, coastal storms, or other currents. They play an important role in building up the continental slopes as well as the smooth ocean bottoms at the foot of the slopes.

Turbidites are sedimentary structures formed by turbidity currents that flow downslope at very high speeds and spread out horizontally, with a broad head up to 2,000 feet wide. Gravels are usually rare in the ocean and are mainly transported from the coast to the deep abyssal plains by turbidity currents. Breccias, or broken, angular rock fragments, are also

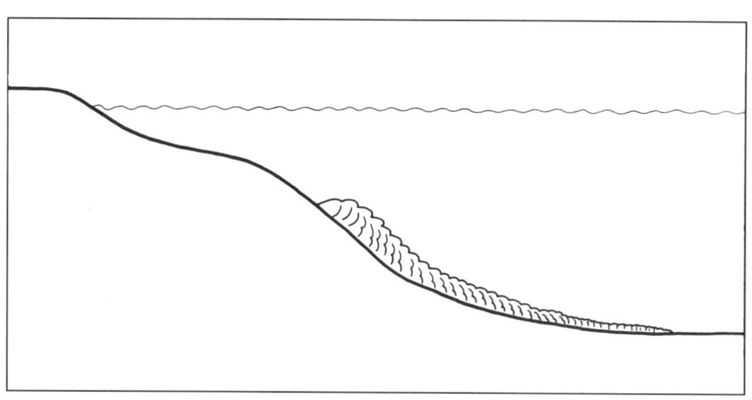

Figure 126 Undersea slides transport large amounts of sediment and carve out deep canyons in the ocean floor.

relatively rare and indicative of terrestrial mudflows or submarine landslides. In addition, debris flows piled up on the continental slope can produce a coarse carbonate rubble known as brecciola.

The continental shelf extends up to 100 miles or more and reaches a depth of roughly 600 feet. In most places, it is nearly flat with an average slope of only about 10 feet per mile. By comparison, the continental slope extends to an average depth of 2 miles or more and has a steeper angle of 2 to 6 degrees, comparable to the slopes of many mountain ranges. Some clifflike submarine slopes might be inclined at very steep angles up to 70 degrees.

Sediments reaching the edge of the continental shelf slide down the continental slope under the influence of gravity. Often, huge masses of sediment cascade down the continental slope by gravity slides that can gouge out steep submarine canyons and deposit great heaps of sediment. They are often just as catastrophic as terrestrial slides and can move massive quantities of sediment downslope in a matter of hours. A modern slide that broke submarine cables near Grand Banks, south of Newfoundland, moved downslope at a rapid speed of about 50 miles per hour, demonstrating the impressive power of water in motion.

9

COLLAPSED STRUCTURES

The surface of the Earth is dimpled by structures resulting from catastrophic collapse. This phenomenon is perhaps best demonstrated at volcanic calderas, formed when the roof of a magma chamber collapses or when a volcano blows off its peak, leaving a broad depression. Large earthquakes whose faults cut the surface also slice up the ground, producing large breaks in the crust called fissures. The dissolution of soluble materials underground or the withdrawal of fluids from subsurface sediments leads to subsidence, or horizontal depressions. Other ground failures occur when subterranean sediments liquify during earthquakes or violent volcanic eruptions, causing the ground to give way.

CALDERAS

Possibly the most explosive volcanic eruption in recorded history occurred during the seventeenth century B.C. on the island of Thera in the Mediterranean Sea, 75 miles north of Crete. The magma chamber beneath the island apparently flooded with seawater, and like a gigantic pressure cooker the volcano blew its lid. The volcanic island collapsed into the emptied magma chamber, forming a deep water-filled caldera that covered an area of 30

Figure 127 The caldera (dashed line) created by the explosive eruption of Krakatau.

square miles. The collapse of Thera also created an immense sea wave that battered the shores of the eastern Mediterranean.

Krakatau lies in the Sunda Strait between Java and Sumatra, Indonesia. On August 27, 1883, a series of four powerful explosions ripped the island apart. The explosions were probably powered by the rapid expansion of steam, generated when seawater entered a breach in the magma chamber. Following the last convulsion, the majority of the island caved into the emptied magma chamber and created a large undersea caldera more than 1,000 feet below sea level (Figure 127), resembling a broken bowl of water with jagged edges protruding above the surface of the sea.

On June 6, 1912, in the Katmai region on the northeast end of the Alaskan Peninsula, a gigantic explosion of unprecedented violence tore open the bottom of the west slope of Mount Katmai. The top 1,200 feet of the volcano exploded and collapsed into a large caldera 1.5 miles wide and 2,000 feet deep that later filled with water to form a crater lake. A mass of 10 cubic miles of pumice, ash, and gas invaded the valley, burning entire forests in its path and filling the valley with volcanic debris, in some places to a height of 600 feet.

The largest volcanic eruption in the continental United States in several centuries occurred on May 18, 1980, when Mount St. Helens exploded with the equivalent force of a 400-megaton nuclear weapon. The eruption blew off the top third of the mountain and created a caldera more than 1 mile wide (Figure 128). It also produced one of the largest avalanches in recorded history and generated massive mudflows and floods that reached all the way to the Pacific Ocean. The blast devastated 200 square miles of forestland, and enough timber to build a fair-size city lay toppled like matchsticks (Figure 129).

Figure 128 A view of Mount St. Helens from the northeast following the May 18, 1980 eruption, showing massive destruction and collapse of the volcano's summit, Skamania County, Washington. Photo by M. M. Brugman, courtesy of USGS

During the last 2 million years, three major episodes of volcanic activity took place in Yellowstone National Park, Wyoming. Some 600,000 years ago, a massive eruption disgorged about 250 cubic miles of ash and pumice, equal to around 1,000 times that of Mount St. Helens. The volcanic eruption created the huge Yellowstone caldera, encompassing an area of about 45 miles long and 25 miles wide. It is counted among the greatest catastrophes of nature, and another major eruption in this area is well overdue. The Yellowstone caldera is a typical resurgent caldera, whose floor has slowly domed upward at an average rate of about three quarters of an inch per year since 1923.

A resurgent caldera forms when the sudden ejection of large volumes of magma from a magma chamber a few miles below the surface abruptly removes the underpinning of the chamber's roof. This collapses the roof, leaving a deep, broad depression on the surface (Figure 130). The infusion of fresh magma into the magma chamber causes a slow upheaval of the floor of the caldera, with a vertical uplift of several hundred feet. If a large part of the caldera floor begins to bulge rapidly at a rate of several feet a day, an eruption usually follows within a few days.

Like all resurgent calderas, the one at Yellowstone formed above a mantle plume, or hot spot, that was long-lasting and large enough to melt huge volumes of rock. Resurgent calderas are recognized by widespread volcanic activity such as hot springs and geysers. The hot spot was not always located beneath Yellowstone, however. Its positions relative to the North American plate can be traced by following volcanic rocks on the Snake River Plain for 400 miles in southern Idaho. Over the past 15 million years, the North American plate slid southwestward across the hot spot, placing

Figure 129 Trees destroyed by the lateral blast from the May 18, 1980, eruption of Mount St. Helens.
Photo by J. Hughes, courtesy of USDA–Forest Service

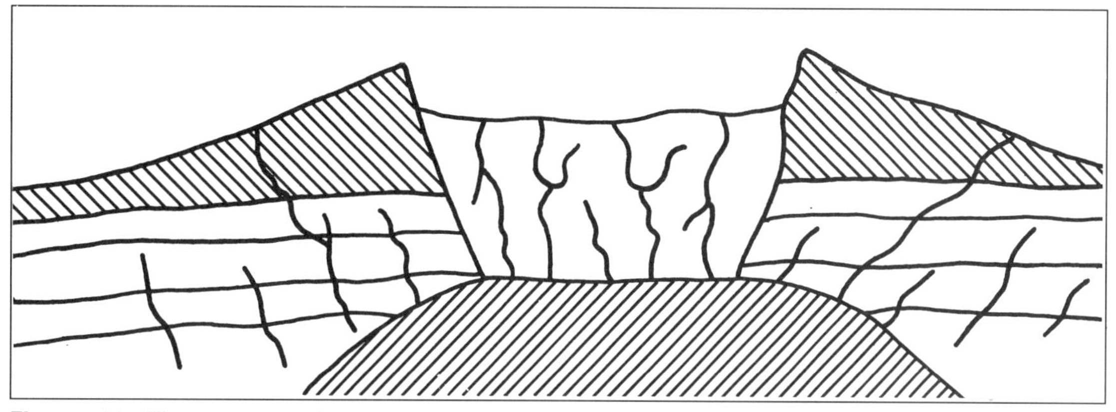

Figure 130 The structure of a resurgent caldera.

it under its temporary home at Yellowstone. Eventually, as the plate continues in its westerly direction, the relative motion of the hot spot will bring its volcanic activity across Wyoming and Montana.

Two other caldera eruptions are known to have occurred in the United States during the last million years. A million years ago, a massive eruption formed the Valles caldera in northern New Mexico. Recently, a deep well was drilled into the hot depths of the dormant volcanic system to test its geothermal energy potential. An injection well was drilled on the flanks of the caldera to a depth of nearly 2 miles and encountered temperatures of 200 degrees Celsius. Water circulating through the porous rocks carried heat away from a magma chamber buried 3 miles below ground. A second recovery well brought the hot water back to the surface (Figure 131).

Figure 131 Steam released from a geothermal well drilled into the Valles caldera in the Jemez Plateau, Sandoval County, New Mexico.
Courtesy of U.S. Department of Energy

Long Valley, California, is a 2-mile-deep depression, resulting from a cataclysmic eruption 700,000 years ago. The caldera is located just east of Yosemite National Park and is about 20 miles long and 10 miles wide. When the volcano erupted, it fragmented the mountains that filled the area into rocky debris. About 140 cubic miles of material were strewn over a wide area, as far away as the East Coast. It appears that magma is once again moving into the resurgent caldera from a depth of 7 miles below the surface. The increased volcanic activity is indicated by a rise in the center of the floor of a foot or more along with numerous medium-size earthquakes since 1980. If this signals an impending eruption, large portions of Nevada could be flooded with thick layers of molten rock.

As many as 10 similar eruptions have occurred in other parts of the world within the last million years. In northern Sumatra, a massive eruption 75,000 years ago created the Toba caldera, which subsided as much as a mile or more. It is the world's largest known resurgent caldera, whose maximum dimension is nearly 60 miles. It is presently filled with a large lake that contains a 25-by-10-mile island created by the upraised floor of the resurgent caldera.

Many other calderas that are no older than 20 to 30 million years lie in a broad belt that covers Nevada, Arizona, Utah, and New Mexico. They generally exist in zones where the crust is thinning, such as rifts, where the mantle rises close to the surface.

Calderas also form in areas where the crust has been fractured, allowing magma to move upward to the surface. The intruded magma domes the overlying crust upward, creating a shallow magma chamber that contains a large volume of molten rock. The doming produces stress in the surface rock that forms the roof of the chamber and causes it to collapse along a ring fracture zone, which becomes the outer wall of the caldera after the eruption.

Smaller calderas are produced when a powerful volcano erupts and decapitates itself by blowing off its upper peak, leaving a broad crater usually over a mile wide. When a volcano erupts both cinder and lava, it builds a composite volcano called a stratovolcano. When the volcano erupts, the hardened plug in its throat is blasted away by the buildup of pressure from trapped gases below. Along with molten rock, fragments are sent aloft and fall back on the volcano's flanks as cinder and ash. The cinder layers are reinforced by layers of lava from milder eruptions, forming cones with a steep summit and steeply sloping flanks.

These processes form the tallest volcanoes, and they often result in a catastrophic collapse. Indeed, massive collapse is what keeps these volcanoes from becoming the highest mountains in the world. For example, Oregon's Mount Mazama collapsed 6,000 years ago, forming a wide caldera that later filled with water and is known today as Crater Lake, the deepest lake in North America.

Isolated volcanic structures, called seamounts, are strung out in chains across the ocean floor. The summit of a seamount sometimes contains a large depression or crater within which lava is extruded (Figure 132). If the crater exceeds 1 mile in diameter it becomes an undersea caldera, whose depth might vary from 150 to 1,000 feet below the rim. Such calderas form when a magma reservoir empties, creating a hollow chamber. Without support, the top of the volcanic cone collapses, creating a wide depression. Feeder vents along the periphery of the caldera supply fresh lava that eventually fills the caldera, resulting in a flattop volcano.

FISSURES

Fissures are large cracks in the crust, resulting from violent earthquakes and volcanic eruptions. Fissure eruptions, in which magma extrudes onto the surface through cracks in the crust rather than a central vent, are the most common type of volcanic eruption. During the Mount Katmai eruption in 1912, powerful explosions ejected millions of tons of volcanic material from a single fissure 5 miles west of the main vent (Figure 133).

Figure 132 The rim of a lava lake collapse pit on the Juan de Fuca Ridge on the East Pacific Rise.
Courtesy of USGS

Figure 133 The Novarupta rhyolite dome is thought to be the main site of the June 6, 1912, Katmai region eruption. Courtesy of USGS

If magma occupies fissures in the crust, it hardens to form tabular intrusive magma bodies called dikes. Their length is considerably longer, upwards of several miles, than their width, which is perhaps only a few feet. Because dike rocks are usually harder than the surrounding material, they generally form long ridges when exposed by erosion. A good example is Ship Rock Peak, New Mexico, where three dikes radiate from a 1,300-foot volcanic neck.

The greatest earthquakes to strike the continental United States in recorded history took place near New Madrid, located in southeastern Missouri on the banks of the Mississippi River. During the winter of 1811–1812 three massive earthquakes with magnitudes ranging upwards of 8.7 struck the region. The town itself was completely demolished when the ground beneath it collapsed from a height of 25 feet to only 12 feet above the level of the river. Deep fissures opened up in the earth, and the ground slid down from bluffs and low hills. Thousands of broken trees fell into the river, and whole sandbars and islands disappeared. The earthquake changed the course of the Mississippi River, and created large lakes in the basins of downdropped crust, the largest of which is the 50-foot-deep Reelfoot Lake (Figure 134).

On June 12, 1897, the Assam region of the Himalayas in northeastern India witnessed an earthquake as powerful as the New Madrid earthquakes, with similar changes in ground level occurring over large areas. The 4,000-foot Assam Hills south of the Brahmaputra River were uplifted about 20 feet. Loose rocks and boulders were tossed high into the air, leaving cavities in the ground where they once stood. Even fence posts came out of their holes. The ground was torn up by deep fissures, and huge clods of dirt were tossed in every direction.

California is well known for its large earthquakes. Fortunately, most have taken place in relatively unpopulated regions. The hamlet of Lone Pine in the Owens Valley, east of the Sierra Nevada Mountains, was destroyed on March 26, 1872, by one of the largest earthquakes in California's history. It opened a deep fissure along a 100-mile line in the Owens Valley. At least 30 people died when their fragile adobe huts collapsed on them. More than 1,000 aftershocks ran through the area during the next three days.

During the 1906 San Francisco earthquake, the ground subsided under buildings, causing them to collapse. Avalanches and landslides occurred in many places. An entire hillside slid down a shallow valley for half a mile. South of Cape Fortunas a hill slid into the sea and created a new cape. Roads and fences crossing the fault were offset horizontally up to 21 feet. Trees were uprooted and fissures appeared in many districts (Figure 135).

The March 27, 1964, Alaskan earthquake that devastated Anchorage and

Figure 134 The lower end of Reelfoot Lake, showing trunks of trees killed by the 1811–1812 New Madrid earthquake submergence in the background. Lake County, Tennessee. Photo by Fuller, courtesy of USGS

other nearby seaports was the largest ever recorded in North America. The earthquake set off landslides, and 30 blocks of Anchorage were destroyed when the city's slippery clay substratum slid toward the sea. Huge fissures opened in the outlying areas, and some of the greatest crustal deformation ever known took place (Figure 136). The area of destruction was estimated at 50,000 square miles, and the earthquake was felt over an area of 500,000 square miles.

Figure 135 Fissures along the San Andreas Fault between Point Reyes Station and Olema, Marin County, California, from the 1906 San Francisco earthquake. Photo by G. K. Gilbert, courtesy of USGS

SUBSIDENCE

Subsidence is the lowering or collapse of the land surface either locally or over broad regional areas due to the withdrawal of fluids and vibrations from earthquakes. Because underground fluids fill intergranular spaces and support sediment grains, the removal of large volumes of fluid such as water or oil results in a loss of grain support, a reduction of intergranular spaces and the compaction of clays. This causes subsurface compaction and subsequent land surface subsidence.

Many parts of the world have been steadily sinking due to the withdrawal of large quantities of groundwater or oil. The most dramatic examples of subsidence are along the Gulf coast of Texas, in Arizona, and in California. Large areas of California's San Joaquin Valley have subsided because of intensive pumping of groundwater. The arid agricultural region is so dependent on groundwater that it accounts for about one fifth of all well water pumped in the United States. The ground is sinking at up to 1 foot per year, and in some places the land has fallen more than 20 feet below former levels.

Figure 136 Wreckage of Government Hill School due to catastrophic subsidence from the March 27, 1964, Alaskan earthquake, Anchorage district, Alaska. Courtesy of USGS

The Houston-Galveston area of Texas has experienced local subsidence as much as 7.5 feet, and an average of 1 foot or more over an area of 2,500 square miles, mostly due to the withdrawal of large amounts of groundwater. In Galveston Bay, subsidence of 3 feet or more occurred over an area of several square miles due to the rapid pumping of oil from the underlying strata. Some coastal towns have subsided so low that they are susceptible to flooding during hurricanes. In Mexico City, overpumping of groundwater has caused some parts of the city to subside at a rate of more than 1 foot per year, often resulting in numerous earth tremors. This might explain why residents ignored the foreshocks that preceded the destructive September 19, 1985, earthquake of 8.1 magnitude that destroyed a large portion of the city.

One of the most dramatic cases of subsidence caused by the withdrawal of oil was at Long Beach, California, where the ground sank to form a huge bowl up to 26 feet deep over an area of 22 square miles. The affected land subsided at the rate of 2 feet per year in some parts of the oil field, and in the downtown area the subsidence amounted to up to 6 feet. The subsidence caused considerable damage (about $100 million) to the city's infrastructure. Most of the subsidence was halted by injecting large amounts of seawater under high pressure back into the underground reservoir. The seawater injection also produced a secondary benefit by increasing the production of the oil field.

Venice, Italy, is slowly drowning because the sea is going up while the city is going down. Much of the subsidence is due to overuse of groundwater, causing the aquifer (water-filled strata) under the city to compact. The cumulative subsidence of Venice over the last 50 years is slightly more than 5 inches. Meanwhile, the Adriatic Sea is rising due to an apparent global warming, resulting in a thermal expansion of the ocean and the melting of the ice caps. The Adriatic has risen about 3.5 inches during this century. Together with the subsidence, that makes for a change of more than 8 inches between Venice and the sea.

Subsidence due to the withdrawal of groundwater can produce fissures or the renewal of surface movement in areas cut by preexisting faults. The fissuring results in the formation of open cracks in the ground. Surface faulting and fissuring resulting from the withdrawal of groundwater is a potential problem in the vicinity of Las Vegas, Nevada, as well as parts of Arizona, California, Texas, and New Mexico. The withdrawal of large

Figure 137 The subsidence of the coast at Halape from the November 29, 1975, Kalapana earthquake, Hawaii County, Hawaii. Photo by R. I. Trilling, courtesy of USGS

volumes of water as well as oil and gas can also cause the ground to subside to considerable depths, sometimes with catastrophic consequences.

Subsidence caused by earthquakes in the United States has taken place mainly in Alaska, California, and Hawaii (Figure 137). The subsidence results from the vertical movement on faults that can affect broad areas. During the 1964 Alaskan earthquake, over 70,000 square miles were tilted downward more than 3 feet, causing extensive flooding. Intense earthquakes can cause subsidence over smaller areas. During the 1811–1812 New Madrid, Missouri, earthquakes, subsurface sand and water were ejected to the surface, leaving voids in the subsurface that caused compaction of subterranean materials and ground settling.

During an earthquake, sand boils often develop in young sedimentary environments where the water table is near the surface. Sand boils are fountains of water and sediment that spout from the pressurized liquefied zone and can reach up to 100 feet high. They are produced when sediment-laden water is vented to the surface by artesian-like water pressures that can excavate large sand pits (Figure 138). Sand boils can also cause local flooding and the accumulation of huge amounts of silt and sand, often in places where they are not wanted. The expulsion of sediment-laden fluids from below ground might also form a large

Figure 138 Sand boils from the October 15, 1979, Imperial Valley earthquake, California. Photo by C. E. Johnson, courtesy of USGS

Figure 139 Craters created by underground nuclear detonations at the Nevada Test Site, Nye County, Nevada. Courtesy of U.S. Department of Energy

subsurface cavity that causes the overlying layers to subside.

Coastal subsidence following an earthquake causes vegetated lowlands that were sufficiently elevated to avoid being inundated by the sea to sink far enough to be submerged regularly and become barren tidal mud flats. Between great earthquakes sediments fill the tidal flats and raise them to the level where vegetation can once again grow. Therefore, repeated earthquakes produce alternating layers of lowland soil and tidal flat mud.

The settling of sediments after the addition of water to the land has caused significant subsidence, especially in the dry western states that have been heavily irrigated, such as the San Joaquin Valley of California. It occurs when dry surface or subsurface deposits are extensively wetted for the first time since their deposition. The wetting causes a reduction in the cohesion between sediment grains, allowing them to move and fill in intergranular openings. This results in the lowering of the land surface from 3 to 6 feet, and as much as 15 feet in the most extreme cases. The effects of the compaction on the land are usually uneven, causing depressions, cracks, and wavy surfaces.

The Nevada Test Site, located 65 miles northwest of Las Vegas, has taken on the appearance of a moonscape, pockmarked with numerous craters created by underground nuclear tests (Figure 139). The craters formed when subterranean sediments fused into glass by the tremendous heat generated by the explosion. This reduced the volume, causing the overlying sediments to collapse to fill in the underground caverns. Sometimes fissures opened on the surface to vent gases escaping from the molten rocks.

Collapse of abandoned underground mines, especially shallow coal mines, occurs often in the eastern United States. The rocks above the mine workings might not have adequate support. When they collapse, the surface drops several feet, forming numerous depressions and pits (Figure 140). Solution mining, which uses water to remove soluble minerals such as salt, gypsum and potash, can produce huge underground cavities, whose collapse causes surface subsidence.

Figure 140 Subsidence depressions, pits and cracks above an abandoned coal mine, Sheridan County, Wyoming. Photo by C.R. Dunrud, courtesy of USGS

Figure 141 Apartment buildings tipped over because of loss of bearing strength caused by liquefaction of a sand layer during the 1964 Niigata earthquake, Japan. Photo by T. L. Youd, courtesy of USGS

The forgotten shafts left in old coal and salt mines might collapse under overlying buildings if not backfilled, but locating the mines can often be difficult. Because an underground cavity has a higher electrical resistance than the surrounding materials, its location and dimensions can be determined by monitoring electrical fields that travel into the ground. The technique might also be useful for discovering underground tunnels and caverns for archaeologists looking for buried artifacts.

CATASTROPHIC COLLAPSE

When the soil supporting buildings or other structures liquefies and loses strength, large deformations can occur within the soil, causing buildings to settle or tip over. Soils that liquefy beneath buildings distort the subsurface geometry, causing bearing failures and subsequent subsidence that can tilt buildings. Normally these deformations occur when a layer of saturated, cohesionless sand or silt extends from near the surface to a depth about equal to the width of the building. The most spectacular example of this type of ground failure occurred during the June 16, 1964, Niigata, Japan, earthquake, when several four-story apartment buildings tilted as much as 60 degrees (Figure 141). The earthquake caused sections of the city to subside a foot or more, resulting in serious flooding when dikes holding back the sea were breached.

Another type of ground failure found in colder climates is called solifluction. When frozen ground melts from the top down, as during warm spring days in the temperate regions or during the summer in permafrost areas, it causes soil to move downslope over a frozen base, damaging any buildings there or even carrying them away.

10

UNUSUAL DEPRESSIONS

The Earth's architecture would not be complete without its many unusual structures created by a variety of geologic processes. Erosion scours the land as well as the seabed, producing deflation basins and a complex seafloor geology. The evacuation of sediment from the land by wind erosion produces blowouts. The expulsion of gases under high pressure also produces blowouts both on land and on the seafloor. The Earth hosts a variety of holes in the ground, including potholes, sinkholes, and numerous craters. Among the most unusual depressions are kimberlite pipes, fumaroles and geysers, crater lakes, and lava lakes. These are just a few of the many great wonders created by the Earth's active geology.

BLOWOUTS

Dust storms often involve a solid wall of dust that blows at up to 60 miles per hour or more (Figure 142). The rising dust clouds can extend several thousand feet high and stretch for hundreds of miles. The land is scoured by the winds, and several inches of soil can be airlifted to other areas. Massive dust storms called haboobs, from an Arabic word meaning "violent wind," arise in the deserts of Africa, Arabia, central Asia, Australia,

TABLE 12 MAJOR DESERTS OF THE WORLD

Desert	Location	Type	Area (square miles × 1,000)
Sahara	North Africa	Tropical	3,500
Australian	Western/interior	Tropical	1,300
Arabian	Arabian Peninsula	Tropical	1,000
Turkestan	Central Asia	Continental	750
North America	S.W. U.S. N. Mexico	Continental	500
Patagonian	Argentiana	Continental	260
Thar	India/Pakistan	Tropical	230
Kalahari	S.W. Africa	Littoral	220
Gobi	Mongolia/China	Continental	200
Takla Makan	Sinkiang, China	Continental	200
Iranian	Iran Afghanistan	Tropical	150
Atacama	Peru/Chile	Littoral	140

and the Americas (Table 12). Enormous dust storms occur when a powerful airstream moves across vast deserts such as those in Africa, where huge dust bands up to 1,500 miles long and 400 miles wide race across the desert floor. Some large African storm systems have been known to carry dust clear across the Atlantic Ocean to South America.

Grains of sand march across the desert under the force of a strong wind by the process of saltation. The sand grains become airborne for a moment, and when they land they kick up other sand grains, which repeat the process (Figure 143). Wind erosion occurs mainly by deflation, which is the removal of large amounts of sediment by windstorms, resulting in a deflation basin. In some areas, deflation produces hollows called blowouts (Figure 144), which are recognized by their typically concave shape. Often after the fine material has been removed, a layer of pebbles remains to protect against further deflation.

Figure 142 A massive dust storm rises over Phoenix, Arizona, on Labor Day 1972. Courtesy of NOAA

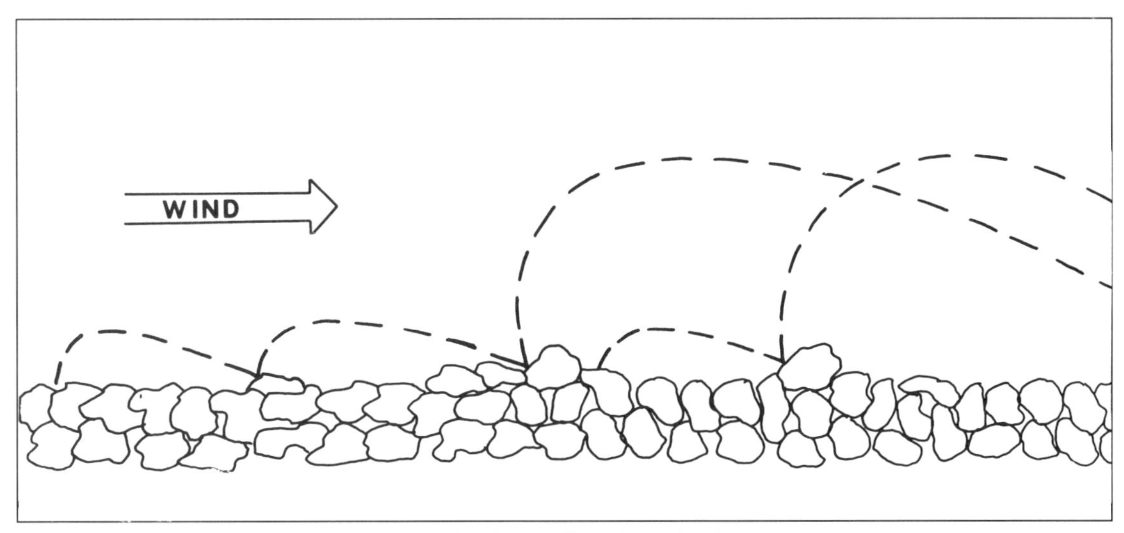

Figure 143 Sand grains march across the desert floor by saltation.

Figure 144 A blowout with core 3 miles south of Harrison, Sioux County, Nebraska. Photo by M.H. Darton, courtesy of USGS

The disruption of the desert in Kuwait, northeast Saudi Arabia, and southern Iraq as a result of the 1991 Persian Gulf war could spawn a new generation of roving sand dunes and a higher incidence of dust storms that sweep the sands from place to place. Military activities during the war have disturbed extensive areas where the natural desert shield has been breached. Over thousands of years, the desert has developed a protective shield of pebbles coated with a desert varnish, ranging in size from a pea to a walnut and too heavy for the strongest desert winds to pick up. This helps hold down sand and dust particles and creates a stable terrain.

Sand dunes generally exist in three basic shapes, determined by topography and patterns of wind flow. Linear dunes, such as those that sweep across the northwest Sahara desert (Figure 145), are aligned in roughly the direction of strong steady prevailing winds. Crescent dunes, also called barchans, are symmetrically shaped with horns pointing downwind. They travel across the desert at speeds of up to 50 feet a year. Star dunes are formed by shifting winds that pile up sand into central points that can rise to 1,500 feet and more (Figure 146).

One curious feature exhibited by sand dunes moving across the desert floor is an unexplained phenomenon known as booming sands. When sand slides down a dune, it sometimes emits a loud, rumbling boom, somewhat like the sound of a jet aircraft flying overhead. The booms can be triggered by simply walking along the dune ridges. The low frequency of the noise appears to originate from some cyclic event occurring at an equally low frequency. However, normal landsliding involves a mass of randomly moving sand grains that collide with a frequency much too high to produce such a boom.

Another type of blowout is produced by the explosive release of gases. Hole in the Ground in the Cascade Mountains of Oregon is the site of a gigantic volcanic gas explosion that left a huge crater. It is a perfectly circular pit that is several thousand feet across and has a rim raised several hundred feet above the surrounding terrain. For unknown reasons, most of the crater is devoid of vegetation.

Similarly, pockets of gases lie trapped under high pressure deep beneath the floor of the ocean. As the pressure increases, the gases explode undersea, spreading debris far and wide and producing huge craters on the ocean floor. The gases rush to the surface in great masses of bubbles that burst in the open air, resulting in a thick foamy froth on the surface of the ocean. A ship sailing into such a foamy sea would immediately lose all buoyancy and sink to the bottom since it is no longer supported by seawater.

Figure 145 Linear dunes crested with barchanoid ridges in the northwest Sahara Desert, Algeria, northern Africa. Photo by E. D. McKee, courtesy of USGS and NASA

In 1906, sailors in the Gulf of Mexico actually witnessed a massive gas blowout that sent mounds of bubbles to the surface. At the site, a large crater was discovered on the ocean floor, lying in 7,000 feet of water southeast of the Mississippi River Delta. The elliptical hole measured 1,300 feet long, 900 feet wide, and 200 feet deep and sat atop a small hill. Downslope laid over 2 million cubic yards of ejected sediment. Apparently, gases seeped upward along cracks in the seafloor and collected under an impermeable barrier. Eventually, the pressure forced the gas to blow off its cover, forming a huge blowout crater.

POTHOLES

Potholes are a common sight on existing river beds as well as on exposed ancient river bottoms (Figure 147). They are generally smooth-sided circular or elliptical holes in hard bedrock such as granite or gneiss, which are

Figure 146 Star domes in the Gran Desierto, Sonora, Mexico. Photo by E. D. McKee, courtesy of USGS

coarse-grained igneous and metamorphic rocks. Potholes have similar shapes, but vary greatly in size, with diameters and depths of up to 5 feet and more. The world's largest pothole is located near Archbald, Pennsylvania, and measures 42 feet wide and nearly 50 feet deep. Huge, rounded boulders lying on the bottom were responsible for its creation, as torrents of water from a melting ice age glacier whirled the rocks around in the hole, widening while simultaneously deepening it.

Another large pothole was formed in the bed of the Deerfield River in Shelburne Falls, Massachusetts, and measures nearly 40 feet across. It is surrounded by several smaller potholes that were cut into hard granite gneiss. Excellent examples of potholes, measuring up to 5 feet in diameter and 30 feet deep, exist on Moss Island in Little Falls, New York. Some of the best potholes are found below dams, which expose the hole-ridden bedrock where rapids once existed.

Most potholes form where water is swiftly flowing and turbulent, such as streams with steep gradients and irregular beds or where a large volume of water is forced through a restricted channel. Such conditions occur when glaciers melt, releasing vast quantities of water that overflow river channels and cause erosion and potholes. They can also form by cavitation caused by the implosion of bubbles in plunge pools below waterfalls. Stream bed slopes tend to steepen where water flows across a boundary between hard rock and soft sediment. The stream erodes the sediment more easily than the hard rock, creating a zone of high-speed turbulent rapids and waterfalls (Figure 148).

Fast, turbulent streams carry rocks that abrade the stream bed as well as each other, becoming rounded in the process. If irregularities or slight depressions occur on the bedrock's surface, water flowing over them is

Figure 147 Potholes in granite ledges in the James River, Henrico County, Virginia. Photo by C. K. Wentworth, courtesy of USGS

Figure 148 The Bridal Veil Falls, Yosemite National Park, Mariposa County, California. Photo by F. E. Matthes, courtesy of USGS

deflected into turbulent eddy currents or whirlpools. Rocks caught in a whirlpool spin around in one place and rub against the sides and bottom of the depression, and abrasion makes the hole wider and deeper. Often, the round, smooth pebbles and boulders that cut the potholes are found on the bottom of the hole, which is often broader than the top. Some potholes that have formed on overhanging ledges might cut completely through the rock to form short, steeply inclined tunnels.

SINKHOLES

Large areas of the country are underlain by limestone and other soluble materials. As groundwater percolates through these sediments, soluble minerals such as calcite dissolve, forming cavities or caverns. The land overlying these caverns can suddenly collapse, forming sinkholes of 300 feet or more in width and 100 feet or more in depth (Figure 149). One of the most dramatic examples of this phenomenon occurred in May 1981, when a sinkhole 350 feet wide and 125 feet deep suddenly opened up in Winter Park, Florida, taking away part of the town.

At other times, the land surface can settle slowly and irregularly. This type of subsidence can cause extensive damage to structures located over pits formed by dissolving soluble minerals. Although the formation of sinkholes is a natural phenomenon, the process can be accelerated by the withdrawal or disposal of water. Sometimes a sinkhole will fill with water and become a small lake.

Figure 149 Possibly the nation's largest sinkhole, which measures 425 feet long, 350 feet wide and 150 feet deep, in central Alabama, Shelby County, Alabama. Courtesy of USGS

The landscape created by such subsidence is called karst terrain. The major locations of karst terrain and caverns are in the southeastern and midwestern United States. They are also found in some portions of the Northeast and West. In Alabama, where soluble limestone and other sediments cover nearly half of the state, thousands of sinkholes pose serious problems for highways and construction. One third of Florida is undergirded by eroded limestone at shallow depths and thus subject to the formation of sinkholes. Throughout the world, some 15 percent of the Earth's surface rests on such terrain, creating millions of sinkholes.

Sinkholes in the shallow waters surrounding the Bahama Islands southeast of Florida are called blue holes because they appear as large dark blue pools of deep water. They formed during the height of the last ice age around 18,000 years ago, when the seas dropped several hundred feet, exposing the area well above sea level. Acid rainwater seeping into the soil dissolved the limestone bedrock and created vast subterranean caverns. Under the weight of the surface rocks, the roofs of the caverns collapsed, exposing huge, gaping pits. The area then refilled with seawater when the great glaciers melted at the end of the ice age. Much fear and superstition surrounds blue holes because they often contain strong eddy currents or whirlpools that can suck a fisherman's boat to the bottom.

Volcanic terrain in parts of Alaska, Washington, Oregon, California, and Hawaii subsides due to local collapse above shallow tunnels in lava flows. Lava tunnels are long caverns beneath the surface of a lava flow created by the withdrawal of lava as the surface hardens. In exceptional cases, they can extend up to 12 miles inside a lava flow. Often, circular or elliptical depressions are found on the surface of the lava flows due to the collapse of the roofs of lava tunnels. A large collapse depression in a lava flow in New Mexico is nearly 1 mile long and 300 feet wide.

SOD PITS

In the fall of 1984, a large chunk of earth in the shape of a keyhole was discovered in an isolated part of north-central Washington State. The area is known as Haystack Rocks for the house-sized boulders, called erratics, that were deposited by retreating glaciers at the end of the ice age. The hunk of sod lay upright and measured 10 feet long, 7 feet wide, and 2 feet thick and weighed about 3 tons. It had vertical sides and a flat bottom as though it had been cut out of the ground by a giant cookie cutter. Indeed, the description is so apt that the slab has been dubbed "the earth cookie."

The piece of turf lay about 73 feet from a hole in the ground with the exact same dimensions. There seemed to be no doubt that the sod came from the hole, which had not been there a month earlier. The grass roots had been ripped out, not cut, as though the plug simply popped out of the ground. A trail of dirt laid in a curved path from the hole to the slab, which was rotated about 20 degrees counterclockwise from the position of the hole from which it came.

No signs of an explosion or any other form of violent activity was found. But the area was hit by a minor earthquake a few days before the plug was discovered. The epicenter was 20 miles away, where the earthquake measured 3.0 in magnitude. However, the distant earthquake did not seem to have enough energy to cause this type of disturbance. Just beneath the topsoil at the site lies a hard layer of bedrock that curves slightly downward to form the shape of a shallow bowl. Perhaps this structure was able to focus the energy of the earthquake in such a way that made the earth jump bodily out of place.

This is not a unique phenomenon, and several similar holes have been discovered in other parts of the world. Earthquakes have been known to toss boulders and people high into the air. Vertical shock waves from a 1797 earthquake in Ecuador supposedly hurled local citizens 100 feet into the air. In 1978, an earthquake in Utah was blamed for creating a depression 2 feet in diameter by tossing fist-sized clods of dirt as far as 14 feet away. One of the most dramatic examples of this phenomenon occurred during the 1897 Assam earthquake in northeast India. The tremor threw huge clods of earth in every direction, some of which landed with their roots pointing toward the sky.

KIMBERLITE PIPES

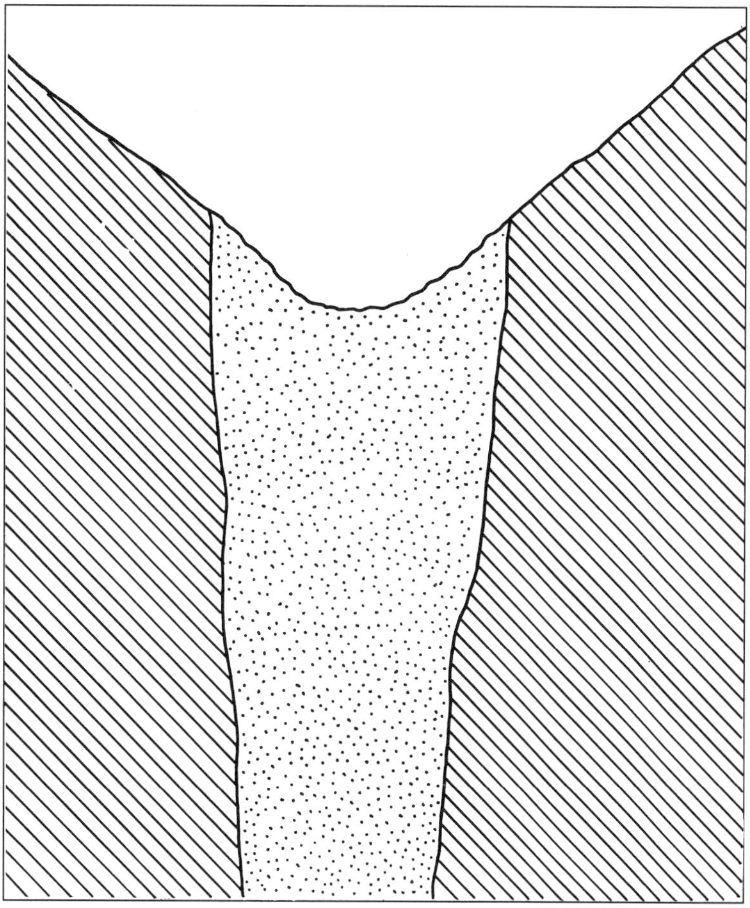

Kimberlite pipes, named for the South African town of Kimberly, are the cores of ancient extinct volcanic structures that extend deep into the upper mantle, as much as 150 miles or more beneath the surface, and have been exposed by erosion. Most known kimberlite pipes were emplaced during the Cretaceous period between 135 and 65 million years ago. They brought diamonds formed during the Archean age from the upper mantle to the surface and are mined extensively for these gems throughout Africa and other parts of the world. Most economic kimberlite pipes are cylindrical or slightly conical structures up to 1 mile across (Figure 150).

Figure 150 Kimberlite pipes, which extend far into the mantle, bring diamonds to the surface.

The pipes are not only important for supplying the world with diamonds, but they also sample a portion of the upper mantle like no other volcanic structure can. At great depth, high temperatures and pressures convert the crystal structure of carbon into a tight lattice, forming the hardest known substance on Earth. Associated with the diamonds are ultramafic nodules that look like rounded cobbles. The nodules were brought up from great depths along with the diamonds and are just as rare. The large proportion of peridotite in the nodules suggests that this mineral comprises a major constituent of the mantle. In addition, xenoliths, from the Greek word meaning "foreign rocks," were among the mantle rocks. They were torn loose from the walls of the volcanic pipe during explosive eruptions, which in turn widened the orifice of the pipe.

The intrusive bodies vary in size and width, and most are roughly circular and pipe-shaped. Over 700 kimberlite pipes and other intrusive structures have been uncovered in South Africa, although only a few contain minable grades of diamonds. The kimberlite deposits were originally worked as open pits, but as the mines grew deeper, underground methods were employed. At the Kimberly mine, the world's deepest diamond mine, the diameter of the pipe at the surface is about 1,000 feet and decreases sharply with depth. Mining discontinued in 1908 at a depth of 3,500 feet due to flooding, even though the diamond-bearing pipe continued on to greater depths.

In North America, a concentration of kimberlite pipes lies along the border between Colorado and Wyoming, and others are known in Montana and in the Canadian Arctic. Few North American pipes are large or economical, with the exception of a solitary pipe near Murfreesboro, Arkansas, which was briefly worked as a diamond mine beginning in 1906 and produced a total of about 40,000 stones (Figure 151). The mine has since been converted to a tourist attraction, known as Crater of Diamonds State Park, where people can sift through black volcanic dirt in search of that illusive stone.

Figure 151 Hydraulic mining at the Arkansas diamond mine in 1923, Pike County, Arkansas. Photo by H. D. Miser, courtesy of USGS

FUMAROLES AND GEYSERS

Alaska's Valley of Ten Thousand Smokes was created when Mount Katmai erupted in June 1912. A series of explosions excavated a depression at the west base of the volcano, whereupon viscous lava 800 feet in diameter rose 195 feet high. The entire valley became a hardened yellowish orange mass 12 miles long and 3 miles wide. Thousands of white fumaroles, which are volcanic steam vents responsible for giving the valley its name, gushed out of the ground and shot hot water vapor up to 1,000 feet into the air.

Fumaroles are vents at the Earth's surface that expel hot gases, usually in volcanic regions. They are found on the surface of lava flows, in the calderas and craters of active volcanoes, and in areas where hot intrusive magma bodies such as plutons occur. The temperature of the gases within a fumarole can reach 1,000 degrees Celsius. Generally, the bulk of the gases consists of steam and carbon dioxide along with smaller quantities of nitrogen, carbon monoxide, argon, hydrogen, and other gases. In other types of fumaroles, called solfataras, from the Italian word meaning "sulfur mine," sulfur gases predominate.

Figure 152 An explosive burst of a geyser at Yellowstone National Park, Wyoming. Photo by D. E. White, courtesy of USGS

The primary requirement for the production of fumaroles and geysers is for a large, slowly cooling magma body to lie close enough to the surface to provide a continuous supply of heat. The hot water and steam are derived either from juvenile water that is released directly from magma melts along with other volatiles or from groundwater that percolates downward near a magma body, where it is heated by convection currents. Volatiles released from the magma body can also heat the groundwater from below.

The term *geyser* comes from the Icelandic word *geysir,* meaning "gusher," which adequately describes its behavior. Geysers are often intermittent and explosive, consisting of hot water ejected with great force, rising 100 to 200 feet. The jet of water is usually followed by a column of thundering steam. The record height was set by New Zealand's Waimangu Geyser, which spouted to 1,500 feet in 1904.

The tube leading to the surface from a deep underground geyser chamber is often restricted or crooked like the drain pipe under a sink. When water seeps into the chamber from a water table, it is heated from the bottom up. The overlying weight of the water places great pressure on the water in the bottom of the chamber, keeping it from boiling, even though temperatures might greatly exceed the normal boiling point of water. As the water temperature gradually increases, some of the water near the top of the geyser tube boils off, decreasing the weight and causing the water in the bottom of the chamber to flash into steam. This overcomes the restriction, and hot water and superheated steam gush out of the vent (Figure 152).

Beneath the huge Yellowstone caldera created by a massive volcanic eruption 600,000 years ago is a volcanic hot spot. It is responsible for the continuous thermal activity that gives rise to a multitude of geysers like Old Faithful, whose nearly hourly eruptions can last 5 minutes and spout a column of steam 130 feet high. In addition, there are a variety of boiling mud pits and hot-water streams, which occur when rainwater seeps into the ground, acquires heat from a magma chamber, and rises through fissures in the torn crust. The region is frequently shaken by earthquakes. The largest in recent times occurred in 1959 and threw off Old Faithful's dependable timing.

CRATER LAKES

If a dormant caldera fills with fresh water from melting snow or rain, it forms a crater lake, whose deepness is determined by the depth of the caldera floor and the level of the water below the rim. Sediment eroded from the wall of the caldera accumulates into thick deposits on the bottom of the lake. The erosion also widens the caldera. Sometimes resurgence of the caldera floor will create an island in the middle of the lake that is capped with young lake sediments.

Crater Lake in Oregon (Figure 153) originated when the upper 5,000 feet of the 12,000-foot composite cone of Mount Mazama collapsed 6,000 years ago and filled with rainwater and melting snow. The lake is 6 miles wide and 2,000 feet deep, the sixth deepest lake in the world. The rim of the caldera rises 500 to 750 feet above the surface of the lake. At one end is a small volcanic peak called Wizard Island, which evolved from later volcanic activity. A similar crater lake was formed when the top 1,200 feet of Alaska's Mount Katmai was blown away by an explosive eruption in June 1912. The eruption created a caldera 1.5 miles wide and 2,000 feet deep that was filled with water from melting snow.

The world's largest crater lake is in northern Sumatra and fills the Toba caldera, which was created by the greatest volcanic eruption in 2 million years. The caldera, which extends for nearly 60 miles in its longest dimension, resulted when the roof collapsed over a large magma chamber 75,000 years ago. The floor of the caldera consequently subsided by over a mile, allowing a deep lake to form. Later, the floor heaved upward several hundred feet like a huge piston. The resurgence of the lake bottom formed a 250-square-mile island in the middle of the lake, called Samosir, which might still be rising.

Lake Nios is a crater lake in Cameroon, Africa, that exploded in August 21, 1986, sending a wall of toxic fumes cascading down the hillside that killed 1,700 people and numerous animals. The disaster might have been triggered by a small earth tremor that cracked the deep lake bottom, releasing volcanic gases under high pressure. This created a huge bubble that burst explosively through the surface of the water, churning the clear blue lake to a murky reddish brown from stirred-up bottom sediments.

Figure 153 Crater Lake National Monument, Klamath County, Oregon. Courtesy of USGS

LAVA LAKES

Lakes on Mount Kilauea, Hawaii, are filled not with water but with molten basalt at 1,200 degrees Celsius (Figure 154). At the summit of most volcanoes is a steep-walled depression called a volcanic crater. The crater is

Figure 154 A lava lake and fountain on Kilauea Volcano created during the 1959–1960 eruption, Hawaii County, Hawaii. Photo by D. H. Richter, courtesy of USGS

connected to the magma chamber by a pipe or vent. When fluid magma moves up the pipe, the lava is stored in the crater until it fills up and overflows. During periods of inactivity, back flow can completely drain the crater.

Kilauea rises a little over three quarters of a mile above sea level and is shaped like an inverted saucer, with a large crater at the summit, from which two rift zones radiate. The eruptions are usually limited to the crater and the rift zones, particularly the eastern rift zone and the fire pit in the crater. On average, Kilauea has erupted at least once a year since 1952. The lava lakes on Kilauea are basalt flows from previous eruptions that have been trapped in large pools and do not solidify to any large extent.

The depth of the lakes can be substantial, as much as 400 feet for the Kilauea Iki Crater. The lakes take a long time to cool and solidify, generally up to one year for a shallow lake to as long as 25 years for the deepest at Kilauea Iki. Eventually, the natural dikes that channel the lava into the lake collapse, and the lake is cut off from its sources and begins to solidify from the bottom up as well as from the top down. Some lava lakes disappear completely down the bottom of the crater as though the drain plug was pulled.

Other lakes often erupt vigorous fire fountains, which are great volumes of lava sprayed high into the air (Figure 155). Mauna Loa is known for its tall fountains of white-hot lava that shoot several hundred feet high, forming a characteristic "curtain of fire." Despite their spectacular and violent outbursts, the eruptions are relatively harmless and a great delight. Their leaps of fire and smell of brimstone is why such volcanos are often referred to as "the gateways to Hell."

Figure 155 An alae eruption of Kilauea Volcano, with a large lava fountain. Photo by G. A. Smathers, courtesy of National Park Service

GLOSSARY

alluvium	stream-deposited sediment
alpine glacier	a mountain glacier or a glacier in a mountain valley
anticline	folded sediments that slope downward away from a central axis
Apollo asteroids	asteroids that come from the main belt between Mars and Jupiter and cross the Earth's orbit
aquifer	a subterranean bed of sediments through which groundwater flows
aragonite	a calcium carbonate mineral similar to calcite and found in cave and hot spring deposits
asteroid	a rocky or metallic body, many of which orbit the sun between Mars and Jupiter, possibly once part of a larger body that subsequently disintegrated
asteroid belt	a band of asteroids orbiting the sun between the orbits of Mars and Jupiter
asthenosphere	a layer of the upper mantle roughly between 50 and 200 miles below the surface that is more plastic than the rock above and below, and might be in convective motion
astrobleme	eroded remains on the Earth's surface of an ancient impact structure produced by a large cosmic body

basalt	a volcanic rock that is dark in color and usually quite fluid in the molten state
batholith	the largest of intrusive igneous bodies and more than 40 square miles on its uppermost surface
bedrock	solid layers of rock lying beneath younger material
black smoker	superheated hydrothermal water rising to the surface at a midocean ridge. The water is supersaturated with metals, and when exiting through the seafloor it quickly cools, and the dissolved metals precipitate, resulting in black, smokelike effluent.
bolide	an exploding meteor whose fireball is often accompanied by a bright light and sound when passing through the Earth's atmosphere
borehole	a hole drilled into the Earth's crust
calcite	a mineral composed of calcium carbonate, and the main constituent of limestone
caldera	a large pitlike depression found at the summits of some volcanoes and formed by great explosive activity and collapse
calving	formation of icebergs by pieces breaking away from glaciers that enter the ocean
carbonaceous chondrites	stony meteorites that contain abundant organic compounds
carbonate	a mineral containing calcium carbonate, such as limestone and dolostone
chondrite	the most common type of meteorite, composed mostly of rocky material with small spherical grains called chondrules
circum-Pacific belt	active seismic regions around the rim of the Pacific plate, coincides with the Ring of Fire
cirque	a glacial erosion feature, producing an amphitheater-like head of a glacial valley
coalescence	the merging of two or more bodies into a single entity

coma	the atmosphere surrounding a comet when it comes within the inner Solar System. The gases and dust particles are blown outward by the solar wind to form the comet's tail.
comet	a celestial body believed to come from the Oort cloud that surrounds the sun and develops a long tail of gas and dust particles when traveling near the inner solar system
continent	a landmass composed of light, granitic rock that rides on denser rocks of the upper mantle
continental drift	the concept that the continents have been drifting across the surface of the Earth throughout geologic time
continental glacier	an ice sheet covering a portion of a continent
continental shelf	the offshore area of a continent in shallow sea
continental shield	ancient crustal rocks upon which the continents grew
continental slope	the transition from the continental margin to the deep-sea basin
convection	a circular, vertical flow of a fluid medium due to heating from below. As materials are heated, they become less dense and rise, while cooler, heavier materials sink.
core	the central part of a planet, consisting of a heavy iron-nickel alloy
correlation	the tracing of equivalent rock exposures over distance
cosmic dust	small meteoroids existing in dust bands possibly created by the disintegration of comets
craton	the stable interior of a continent, usually composed of the oldest rocks on the continent
crevasse	a deep fissure in the Earth or a glacier
crust	the outer layers of a planet's or a moon's rocks
crustal plate	one of several plates comprising the Earth's surface rocks

desiccated basin	a basin formed when an ancient sea evaporated
diapir	the buoyant rise of molten rock through heavier rock
differentiation	the separation of solids or liquids according to their weight with heavy masses sinking and light material rising toward the surface
dike	a body of intrusive igneous rock that cuts across the layering or structural fabric of the host rock
divergent plate boundary	the boundary between crustal plates where the plates move apart. Generally corresponds to the midocean ridges where new crust is formed by the solidification of liquid rock rising from below.
dolostone	a carbonate rock formed by the replacement of calcium with magnesium in limestone
domepit	a vertical shaft connecting two cave passages
dropstone	a large boulder thought to be deposited on an ancient seabed after being rafted out to sea by glacial ice
drumlin	a hill of glacial debris facing in the direction of glacial movement
dune	a ridge of windblown sediments usually in motion
earthquake	the sudden breaking of the Earth's rocks
East Pacific Rise	a midocean spreading center that runs north-south along the eastern side of the Pacific. The predominant location upon which the hot springs and black smokers have been discovered.
eolian	describing a deposit of windblown sediment
erratic	a glacially deposited boulder far from its source
escarpment	a mountain wall caused by elevation of a block of land
esker	curved ridges of glacially deposited material
evaporite	the deposition of salt, anhydrite, and gypsum from the evaporation of stranded seawater in an enclosed basin
extrusive	any igneous volcanic rock ejected onto the surface of a planet or moon

fault	a breaking of crustal rocks caused by earth movements
fissure	a large crack in the crust through which magma might escape
flowstone	a mineral deposit formed on the walls and floor of a cave
fossil	any remains, impression, or trace in rock of a plant or animal of a previous geologic age
formation	a combination of rock units that can be traced over distance
fractionation	a process by which a subducted slab of crustal rock starts melting on its way down into the mantle, while the lighter components rise back to the surface
frost heaving	the lifting of rocks to the surface by the expansion of freezing water
frost polygons	polygonal patterns of rocks formed by repeated freezing
fumarole	a vent through which steam or other hot gases escape from underground, such as a geyser
geothermal	the generation of hot water or steam by hot rocks in the Earth's interior
geyser	a spring that ejects intermittent jets of steam and hot water
glacier	a thick mass of moving ice occurring where winter snowfall exceeds summer melting
gneiss	a banded, coarse-grained metamorphic rock with alternating layers of unlike minerals, consisting of essentially the same components as granite
Gondwana	a southern supercontinent of Paleozoic time, consisting of Africa, South America, India, Australia, and Antarctica. It broke up into present continents during the Mesozoic era.
graben	a valley formed by a downdropped fault block
granite	a coarse-grained, silica-rich rock consisting primarily of quartz and feldspars. It is the principal constituent of

	the continents and is believed to evolve from a molten state beneath the Earth's surface.
groundwater	the water derived from the atmosphere that percolates and circulates below the surface of the Earth
guyot	an undersea volcano that reached the surface of the ocean, whereupon its top was flattened by erosion. Later subsidence caused the volcano to sink below the surface, preserving its flattop appearance.
gypsum	an evaporite mineral composed of calcium sulfate
half-life	the time for half the atoms of a radioactive element to decay
helictite	a branching calcium carbonate deposit on cave walls
horn	a peak on a mountain formed by glacial erosion
horst	an elongated, uplifted block of crust bounded by faults
hot spot	a volcanic center that has no relation to a plate boundary location; an anomalous magma generation site in the mantle
hydrothermal	relating to the movement of hot water through the crust
hypocenter	the point of origin of earthquakes; also called the focus
ice age	a period of time when large areas of the Earth were covered by glaciers
iceberg	a portion of a glacier broken off upon entering the sea
ice cap	a polar cover of ice and snow
igneous rocks	all rocks that have solidified from a molten state
impact	the point on the surface upon which a celestial object lands
interglacial	a warming period between glacial periods
intrusive	any igneous body that has solidified in place below the surface of the Earth
iridium	a rare isotope of platinum, relatively abundant on meteorites

island arc	volcanoes landward of a subduction zone that parallel a trench, and exist above the melting zone of a subducting plate
isostasy	the geologic principle that states the crust of the Earth is buoyant and rises or sinks depending on it weight
isotope	a species of an element with the same number of protons but a different number of neutrons in the nucleus
karst	a terrain consisting of numerous sinkholes in limestone
kettle	a depression in the ground caused by a buried block of glacial ice
Kirkwood gaps	bands in the asteroid belt that are mostly empty of asteroids due to Jupiter's gravitational attraction
lahar	a hot mudflow or ashflow on the slopes of a volcano
lamellae	striations on the surface of crystals caused by a sudden release of high pressures, such as those created by large meteorite impacts
landslide	rapid downhill movement of Earth materials often triggered by earthquakes
Laurasia	the northern supercontinent of the Paleozoic, consisting of North America, Europe, and Asia
lava	molten magma after it has flowed out onto the surface
limestone	a sedimentary rock consisting mostly of calcite
liquefaction	the liquefying of sediment layers generally due to earthquake activity
lithosphere	a rigid outer layer of the mantle, typically about 60 miles thick. It is overridden by the continental and oceanic crusts and is divided into segments called plates.
loess	a thick deposit of airborne dust
magma	a molten rock material generated within the Earth that is the constituent of igneous rocks, often extruded by volcanic eruptions

mantle	the part of a planet below the crust and above the core, composed of dense iron-magnesium-rich rocks
maria	dark plains on the lunar surface caused by massive basalt floods
mass wasting	the downslope movement of rock under the direct influence of gravity
megaplume	a large volume of mineral-rich warm water above an oceanic rift
metamorphic rock	a rock crystallized from previous igneous, metamorphic, or sedimentary rocks created under conditions of intense temperatures and pressures without melting
meteor	a small celestial body that becomes visible as a streak of light when entering the Earth's atmosphere
meteorite	a metallic or stony body from space that enters the Earth's atmosphere and impacts on the Earth's surface
meteoritic crater	a depression in the crust produced by the bombardment of a large meteorite
meteoritics	the science that deals with meteors and related phenomena
meteoroid	general term for a meteor in orbit around the sun. Compare with meteor, which is a meteoroid that has entered the Earth's atmosphere.
meteor shower	a phenomenon observed when large numbers of meteors enter the Earth's atmosphere. Their luminous paths appear to diverge from a single point.
micrometeorites	small, grain-size bodies that strike spacecraft
microtektites	small, spherical grains created by the melting of surface rocks during a large meteorite impact
Mid-Atlantic Ridge	the seafloor spreading ridge of volcanoes that marks the extensional edge of the North American and South American plates to the west and the Eurasian and African plates to the east.

midocean ridge	a submarine ridge along a divergent plate boundary where a new ocean floor is created by the upwelling of mantle material
moraine	a ridge of erosional debris deposited by the melting margin of a glacier
morphology	the general shape of a geologic structure
mountain roots	the deeper crustal layers under mountains
Oort Cloud	the collection of comets that surround the sun about a light-year away
orogeny	a process of mountain building by tectonic activity
outgassing	the loss of gas within a planet as opposed to degassing, or loss of gas from meteorites
Pangaea	an ancient supercontinent that included all the lands of the Earth
Panthalassa	the great world ocean that surrounded Pangaea
permafrost	permanently frozen ground in the arctic regions
permeability	the ability to transfer fluid through cracks, pores, and interconnected spaces within a rock
placer	a deposit of rocks left behind from a melting glacier; any ore deposit enriched by stream action
planetesimal	a small celestial body that might have existed in the early stage of the Solar System
planetoid	a small body, generally no larger than the moon, in orbit around the sun. A disintegration of several such bodies might have been responsible for the asteroid belt between Mars and Jupiter.
plate tectonics	the theory that accounts for the major features of the Earth's surface in terms of the interaction of lithospheric plates
pluton	an underground body of igneous rock younger than the rocks that surround it. It is formed where molten rock oozes into a space between older rocks.

porosity — the percentage of pore spaces in a rock between crystals and grains

radiometric dating — a method of determining the age of an object by chemical analysis of stable versus unstable radioactive elements

radionuclide — a radioactive element responsible for generating the Earth's internal heat

reef — the biological community that lives at the edge of an island or continent. The shells form a limestone deposit that is readily preserved in the geologic record.

regression — a fall in sea level, exposing continental shelves to erosion

resurgent caldera — a large caldera that experiences renewed volcanic activity that domes up the caldera floor

rift valley — the center of an extensional spreading center where continental or oceanic plate separation occurs

rille — a trench formed by a collapsed lava tunnel

saltation — the movement of sand grains by wind or water

sandstone — a sedimentary rock consisting of sand grains cemented together

scarp — a steep slope formed by earth movements

schist — a finely layered metamorphic rock that tends to split readily into thin flakes

seafloor spreading — a theory that maintains the ocean floor is created by the separation of lithospheric plates along the midocean ridges, with new oceanic crust formed from mantle material that rises from the mantle to fill the rift

seamount — a submarine volcano

sedimentary rock — a rock composed of fragments cemented together

seismic sea wave — an ocean wave related to an undersea earthquake

shield — areas of the exposed Precambrian nucleus of a continent

shield volcano — a broad, low-lying volcanic cone built up by lava flows of low viscosity

sinkhole	a large pit formed by the collapse of surface materials undercut by the dissolution of subterranean limestone
spherules	small, spherical, glassy grains found on certain types of meteorites, lunar soils, and at large meteorite impact sites on Earth
stalactite	a calcium carbonate cave deposit that grows downward from the ceiling
stalagmite	a calcium carbonate cave deposit that grows upward from the floor
stishovite	a quartz mineral produced by extremely high pressures, such as those generated by a large meteorite impact
stratovolcano	an intermediate volcano characterized by a stratified structure from alternating emissions of lava and fragments
strewn field	a usually large area where tektites are found arising from a large meteorite impact
striation	scratches on bedrock made by rocks embedded in a moving glacier
stromatolite	a calcareous structure built by successive layers of bacteria and which are up to 3.5 billion years old
subduction zone	an area where the oceanic plate dives below a continental plate into the asthenosphere. Ocean trenches are the surface expression of a subduction zone.
subsidence	the collapse of sediments due to the removal of underground fluids
supernova	an enormous stellar explosion in which all but the inner core of a star is blown off into interstellar space, producing as much energy in a few days as the sun does in a billion years.
surge glacier	a continental glacier that heads toward the sea at a high rate of advance
syncline	a fold in which the beds slope inward toward a common axis

tectonic activity	the formation of the Earth's crust by large-scale earth movements throughout geologic time
tektites	small, glassy minerals created from the melting of surface rocks by an impact of a large meteorite
Tethys Sea	the hypothetical mid-latitude area of the oceans separating the northern and southern continents of Gondwana and Laurasia hundreds of millions of years ago
transform fault	a fracture in the Earth's crust along which lateral movement occurs. They are common features of the midocean ridges.
transgression	a rise in sea level that causes flooding of the shallow edges of continental margins
travertine	a calcium carbonate mineral found in cave and hot-spring deposits
trench	a depression on the ocean floor caused by subduction
tundra	permanently frozen ground at high latitudes and high altitudes
uniformitarianism	the theory that the slow processes that shape the Earth's surface have acted essentially unchanged throughout geologic time
varve	thinly laminated lake bed sediments
volcanic cone	the general term applied to any volcanic mountain with a conical shape
volcanic crater	the inverted conical depression found at the summit of most volcanoes and formed by the explosive emission of volcanic ejecta

BIBLIOGRAPHY

Meteorite Impacts

Alvarez, Walter, and Frank Asaro. "An Extraterrestrial Impact." *Scientific American* 263 (October 1990): 78–84.

Alvarez, Luis W. "Mass Extinctions Caused by Large Bolide Impacts." *Physics Today* 40 (July 1987): 24–33.

Boss, Alan P. "The Origin of the Moon." *Science* 231 (January 24, 1986): 341–345.

Gehrels, Tom. "Asteroids and Comets." *Physics Today* 38 (February 1985): 33–41.

Jones, Richard C., and Anthony N. Stranges. "Unraveling Origins, the Archean." *Earth Science* 42 (Winter 1989): 20–22.

Lunine, J.I. "Origin and Evolution of Outer Solar System Atmospheres." *Science* 245 (July 14, 1989): 141–146.

Morrison, David. "Target Earth: It Will Happen." *Sky & Telescope* 79 (March 1990): 261–265.

Newsom, Horton E., and Kenneth W.W. Sims. "Core Formation During Early Accretion of the Earth." *Science* 252 (May 17, 1991): 926–933.

Weissman, Paul R. "Are Periodic Bombardments Real?" *Sky & Telescope* 79 (March 1990): 266–270.

Wetherill, George W. "The Formation of the Earth from Planetesimals." *Scientific American* 224 (January 1981): 163–174.

Impact Craters

Grieve, Richard A. F. "Impact Cratering on the Earth." *Scientific American* 262 (April 1990): 66–73.

Hildebrand, Alan R., and William V. Boynton. "Cretaceous Ground Zero." *Natural History* (June 1991): 47–52.

Lowe, Donald R., et al. "Geological and Geochemical Record of 3400-Million-Year-Old Terrestrial Meteorite Impacts." *Science* 245 (September 1, 1989): 959–962.

Monastersky, Richard. "You Just Can't Wear Them Down." *Science News* 132 (November 7, 1987): 301.

O'Keefe, John A. "The Tektite Problem." *Scientific American* 239 (August 1978): 116–125.

Perth, Nigel Henbest. "Meteorite Bonanza in Australian Desert." *New Scientist* (April 20, 1991): 20.

Sharpton, Virgil L. "Glasses Sharpen Impact Views." *Geotimes* 33 (June 1988): 10–11.

Simon, Cheryl. "Deep Crust Hints Meteoritic Impact." *Science News* 121 (January 30, 1982): 69.

Weisburd, Stefi. "Traces of the Oldest Meteorite Impact?" *Science News* 129 (February 1, 1986): 69.

Wetherill, George W. "Occurrence of Giant Impacts During the Growth of the Terrestrial Planets." *Science* 228 (May 17, 1985): 877–879.

CAVES AND CAVERNS

Bahn, Paul G. "Ice Age Drawings on Open Rock Faces in the Pyrenees." *Nature* 313 (February 14, 1985): 530–531.

Beard, Jonathan. "Glaciers on the Run." *Science 85* 6 (February 1985): 84.

Bolton, David W. "Underground Frontiers." *Earth Science* 40 (Summer 1987): 16–18.

Bower, Bruce. "Cave Evidence Chews up Cannibalism Claims." *Science News* 139 (June 1, 1991): 341.

Hansen, Michael C. "Ohio Natural Bridges." *Earth Science* 41 (Winter 1988): 10–12.

Lipske, Mike. "Wonder Holes." *International Wildlife* 20 (February 1990): 47–51.

Mathews, Samuel W. "Ice on the World." *National Geographic* 171 (January 1987): 84–103.

Mollenhauer, Eric, and George Bartunek. "Glacier on the Move." *Earth Science* 41 (Spring 1988): 21–24.

Petrini, Cathy. "Heart of the Mountain." *Earth Science* 41 (Summer 1988): 14–18.

Canyons and Valleys

Bartusiak, Marcia. "Mapping the Sea Floor from Space." *Popular Science* 234 (February 1984): 81–85.

Bernardo, Stephanie. "The Seafloor: A Clear View from Space." *Science Digest* 92 (June 1984): 44–48.

Birnbaum, Stephen. "The Grand Canyon." *Good Housekeeping* (November 1991): 150–152.

Bower, Bruce. "Shuttle Radar Is Key to Sahara's Secrets." *Science News* 125 (April 21, 1984): 244.

Francheteau, Jean. "The Oceanic Crust." *Scientific American* 249 (September 1983): 114–129.

Monastersky, Richard. "What's New in the Ol' Grand?" *Science News* 132 (December 19 & 26, 1987): 392–395.

Prestrong, Ray. "It's about Time." *Earth Science* 42 (Summer 1989): 14–15.

Radok, Uwe. "The Antarctic Ice." *Scientific American* 253 (August 1985): 98–106.

Weisburd, Stefi. "Sea-Surface Shape by Satellite." *Science News* 129 (January 18, 1986): 37.

Major Basins

Bird, Peter. "Formation of the Rocky Mountains, Western United States: A Continuum Computer Model." *Science* 239 (March 23, 1988): 1501–1507.

Dietz, Robert S., and Mitchell Woodhouse. "Mediterranean Theory May Be All Wet." *Geotimes* 33 (May 1988): 4.

Hekinian, Roger. "Undersea Volcanoes." *Scientific American* 251 (July 1984): 46–55.

Hsu, Kenneth J. "When the Black Sea Was Drained." *Scientific American* 238 (May 1978): 53–63.

———. "When the Mediterranean Dried Up." *Scientific American* 227 (December 1972): 27–36.

Jordan, Thomas H., and J. Bernard Minster. "Measuring Crustal Deformation in the American West." *Scientific American* 259 (August 1988): 48–58.

Kerr, Richard A. "Making Mountains with Lithospheric Drips." *Science* 239 (February 26, 1988): 978–979.

Kunzig, Robert. "Birth of a Nation." *Discover* 11 (February 1990): 26–27.

Johnston, Arch C. "A Major Earthquake Zone on the Mississippi." *Scientific American* 246 (April 1982): 60–68.

Steinhorn, Ilana, and Joel R. Gat. "The Dead Sea." *Scientific American* 249 (October 1983): 102–109.

Volcanic Rifts

Bonatti, Enrico. "The Rifting of Continents." *Scientific American* 256 (March 1987): 97–103.

Courtillot, Vincent, and Gregory E. Vink. "How Continents Break Up." *Scientific American* 249 (July 1983): 43–49.

Gurnis, Michael. "Ridge Spreading, Subduction, and Sea Level Fluctuations." *Science* 250 (November 16, 1990): 970–972.

Macdonald, Kenneth C., and Paul J. Fox. "The Mid-Ocean Ridge." *Scientific American* 262 (June 1990): 72–79.

Monastersky, Richard. "Spinning the Supercontinent Cycle." *Science News* 135 (June 3, 1989): 344–346.

Powell, Corey S. "Peering Inward." *Scientific American* 264 (June 1991): 100–111.

Rampino, Michael R., and Richard B. Strothers. "Flood Basalt Volcanism During the Past 250 Million Years." *Science* 241 (August 5, 1988): 663–667.

Stater, Curt. "Africa's Great Rift." *National Geographic* 177 (May 1990): 10–41.

Tazieff, Haaroun. "The Afar Triangle." *Scientific American* 222 (February 1970): 32–40.

White, Robert S., and Dan P. McKenzie. "Volcanism at Rifts." *Scientific American* 261 (July 1989): 62–71.

Earthquake Faults

Cook, Frederick A., Larry D. Brown, and Jack E. Oliver. "The Southern Appalachians and the Growth of the Continents." *Scientific American* 243 (October 1980): 156–168.

Ellsworth, William L. "Putting the Pieces Together." *Nature* 349 (January 31, 1991): 371–372.

Frohlich, Cliff. "Deep Earthquakes." *Scientific American* 260 (January 1989): 48–55.

Heaton, Thomas H. "Earthquake Hazards on the Cascadia Subduction Zone." *Science* 236 (April 10, 1987): 162–168.

Johnston, Arch C., and Lisa R. Kanter. "Earthquakes in Stable Continental Crust." *Scientific American* 262 (March 1990): 68–75.

Monastersky, Richard. "Tibet's Tectonic Escape Act." *Science News* 138 (July 14, 1990): 24–25.

Roman, Mark B. "Finding Fault." *Discover* 9 (August 1988): 57–70.

Stein, Ross S., and Robert S. Yeats. "Hidden Earthquakes." *Scientific American* 260 (June 1989): 48–57.

Ground Failures

"Facing Geologic and Hydrologic Hazards." *U.S. Geological Survey Professional Paper* 1240-B. Government Printing Office. 1981.

Friedman, Gerald M. "Slides and Slumps." *Earth Science* 41 (Fall 1988): 21–23.

Friend, P. F. "Storms in the Abyss." *Nature* 301 (May 17, 1984): 212.

Hollister, Charles D., Arthur R. M. Nowell, and Peter A. Jumars. "The Dynamic Abyss." *Scientific American* 250 (March 1984): 42–53.

McCave, I. N. "Hummocky Sand Deposits Generated by Storms at Sea." *Nature* 313 (February 14, 1985): 533.

Monastersky, Richard. "Soil May Signal Imminent Landslide." *Science News* 134 (November 12, 1988): 318.

Norris, Robert M. "Sea Cliff Erosion." *Geotimes* 35 (November 1990): 16–17.

Shaefer, Stephen J., and Stanley N. Williams. "Landslide Hazards." *Geotimes* 36 (May 1991): 20–22.

Zimmer, Carl. "Landslide Victory." *Discover* 12 (February 1991): 66–69.

Collapsed Structures

Decker, Robert, and Barbara Decker. "The Eruptions of Mount St. Helens." *Scientific American* 244 (March 1981): 68–80.

Francis, Peter. "Giant Volcanic Calderas." *Scientific American* 248 (June 1983): 60–70.

Francis, Peter, and Stephen Self. "Collapsing Volcanoes." *Scientific American* 256 (June 1987): 91–97.

Kerr, Richard A. "Drilling into Surprises Beneath an Inyo Crater." *Science* 239 (January 22, 1988): 350–351.

Marsden, Sullivan S., Jr., and Stanley N. Davis. "Geological Subsidence." *Scientific American* 216 (June 1967): 93–100.

Simon, Cheryl. "A Giant's Troubled Sleep." *Science News* 124 (July 16, 1983): 40–41.

Unklesbay, A. G. "Midwest Earthquakes." *Earth Science* 40 (Winter 1987): 11–13.

Weisburd, Stefi. "Sensing the Voids Underground." *Science News* 130 (November 22, 1986): 329.

Unusual Depressions

Cook, Patrick. "The Movable Earth Puzzle." *Science 86* 7 (May 1986): 80.

Cox, Keith G. "Kimberlite Pipes." *Scientific American* 238 (April 1978): 120-132.

Edmond, John M., and Karen Von Damm. "Hot Springs on the Ocean Floor." *Scientific American* 248 (April 1983): 78-93.

Goodwin, Bruce K. "The Hole Truth." *Earth Science* 41 (Summer 1988): 23-25.

Holden, Constance. "Kuwait's Unjust Deserts: Damage to Its Desert." *Science* 251 (March 8, 1991): 1175.

Idso, Sherwood B. "Dust Storms." *Scientific American* 235 (October 1976): 108–114.

Kerr, Richard A. "Surging Plate Tectonics Erupts Beneath the Sea." *Science* 250 (December 21, 1990): 1661.

Peck, Dallas L., Thomas L. Wright, and Robert Decker. "The Lava Lakes of Kilauea." *Scientific American* 241 (October 1979): 114–128.

INDEX

N

O

P

Q

R

S